Dopamine Receptors

Dopamine Receptors

Carl Kaiser, EDITOR
Smith Kline & French Laboratories

John W. Kebabian, EDITOR
National Institutes of Health

Based on a symposium sponsored
by the ACS Division of
Medicinal Chemistry
at the 184th Meeting
of the American Chemical Society
Kansas City, Missouri,
September 12–17, 1982

ACS SYMPOSIUM SERIES 224

AMERICAN CHEMICAL SOCIETY

WASHINGTON, D.C. 1983

Library of Congress Cataloging in Publication Data

Dopamine receptors.
(ACS Symposium series, ISSN 0097-6156; 224)

"Based on a symposium sponsored by the ACS
Division of Medicinal Chemistry at the 184th meet-
ing of the American Chemical Society, Kansas City,
Missouri, September 12–17, 1982."
Bibliography: p.
Includes index.
1. Dopamine—Receptors—Congresses. 2. Dopa-
mine—Agonists—Congresses.
I. Kaiser, Carl, 1929–   . II. Kebabian, J. W.
(John W.) III. American Chemical Society. National
Meeting (184th: 1982: Kansas City, Mo.) IV.
Series. [DNLM: 1. Receptors, Dopamine—Congresses.
2. Receptors, Dopamine—Drug effects—Congresses.
WL 102.8 D692 1982]
QP563.D66D66   1983   612'.015   83–6433
ISBN 0–8412–0781–X

ACS Symposium Series

M. Joan Comstock, *Series Editor*

FOREWORD

The ACS SYMPOSIUM SERIES was founded in 1974 to provide
a medium for publishing symposia quickly in book form. The
format of the Series parallels that of the continuing ADVANCES
IN CHEMISTRY SERIES except that in order to save time the
papers are not typeset but are reproduced as they are submitted by the authors in camera-ready form. Papers are reviewed under the supervision of the Editors with the assistance
of the Series Advisory Board and are selected to maintain the
integrity of the symposia; however, verbatim reproductions of
previously published papers are not accepted. Both reviews
and reports of research are acceptable since symposia may
embrace both types of presentation.

CONTENTS

EDITORS' PREFACE

T HE INVITATION BY J. L. NEUMEYER and the Division of Medicinal Chemistry of the American Chemical Society to organize the symposia "Multiple Categories of Dopamine Receptors" and "Modulation of Dopamine Receptors" offered us the opportunity to highlight some of the recent advances in the understanding of dopamine receptors and the drugs interacting with these receptors. Several years ago, John Kebabian presented the "two dopamine receptor hypothesis." This hypothesis is the theme of the first symposium. Similarly, several years ago, Carl Kaiser participated in the discovery of SK&F 38393, a dopaminergic agonist relatively selective for the D-1 receptor. The second symposium focuses attention on the development of novel agonists for dopamine receptors and their use as therapeutic agents. Because some of these differentiate between different dopamine receptors, the concept of multiple categories of dopamine receptors is an integral part of the second symposium.

The requirement of the American Chemical Society that all material published by the Society be subjected to outside review offered an opportunity for us to solicit a second (and sometimes conflicting) opinion about the material presented in these symposia. It cannot be denied that the topic of dopamine receptor(s) remains an area of ongoing investigation, controversy, and disagreement. For many questions about dopamine receptors, the "final verdict" is not yet in. Indeed, if these symposia have delineated the areas of disagreement, the readers of this volume form the jury. For each disagreement or misunderstanding highlighted in this volume, the reader can review the ideas of the different authors and then decide which observations and interpretations are most helpful to their individual endeavors.

The editors are grateful to each of the authors of chapters and to those who contributed commentaries on these chapters. We also acknowledge with gratitude the financial support of Smith Kline & French Laboratories.

CARL KAISER
Smith Kline & French Laboratories
Philadelphia, PA 19101

JOHN W. KEBABIAN
National Institutes of Health
Bethesda, MD 20205

February 1983

PREFACE

BY LESLIE L. IVERSEN

THE CHAPTERS PRESENTED HERE REPRESENT an interesting state-of-the-art reflection of current trends in research on dopamine receptors. D. Calne and T. A. Larsen are probably accurate in suggesting that "dopamine has overtaken acetylcholine and norepinephrine as the most extensively investigated neurotransmitter in the nervous system." Their admirable survey of present and potential clinical uses of dopamine agonists and antagonists indicates that this research effort has indeed led to useful new therapeutic agents, including those acting on peripheral or endocrine targets as well as centrally acting drugs. In particular, the cardiovascular and endocrine effects of dopamine have attracted considerable interest, both from the basic and applied research viewpoints. Furthermore, S. Szabo and J. L. Neumeyer suggest that the actions of dopamine in the gastrointestinal tract, and its possible etiological role in peptic ulcers, will represent an important new focus for future work on the peripheral actions of dopamine.

The availability of peripheral models offers an important means of characterizing the pharmacological properties of dopamine receptors. In many cases it is possible to measure a clear-cut tissue response and, thus, to establish the agonist, partial agonist, and antagonist properties of test compounds. It is perhaps only now being recognized by neurochemists that receptors cannot be characterized fully in any other way. Ten years ago, with the discovery of new biochemical approaches to the study of dopamine receptors in CNS many of us were doubtless too optimistic in thinking that such approaches would lead to rapid progress in defining the characteristics of dopamine receptors. The fundamental problem in achieving such understanding, however, has been that we do not know what dopamine does as neurotransmitter in the various CNS pathways that contain it. Furthermore, there are no simple model systems that allow one to measure agonist and antagonist effects on CNS targets. We cannot know how reliable the results of neurochemical studies of dopamine receptors are if we have no biological response against which to assess the neurochemical data. A recent review (1) listed some 28 different applications of radioligand binding methods to the study of dopamine receptors in brain, that use more than twenty different agonist or antagonist radioligands. Remarkably little is said in the present volume about radioligand binding assays, and I suspect J. C. Stoof caught the mood of the meeting in stating: "Binding studies, although easy to perform, have yielded too many data, too many categories of dopamine receptors and too many controversies." The neurochemical emphasis has clearly shifted to "functional" assays, in which some biochemical response is measured, rather than simply occupation of receptor binding sites by ligands. The activation or inhibition of adenylate cyclase has proved a valuable model in this sense, and other biochemical responses may prove similarly useful (e.g., inhibition of peptide hormone or neurotransmitter release in response to dopamine agonists).

In this volume a good deal of emphasis is placed on studies of dopamine receptors in pituitary. This emphasis seems well justified. The clear-cut effects of dopamine in suppressing prolactin secretion from anterior lobe mammotrophs, and the inhibitory effects on secretion of α-MSH and related secretory products from intermediate lobe are important models for dopamine receptor studies. In both cases new evidence was put forward to support the hypothesis that the actions of dopamine on the secretory cells are mediated by inhibition of adenylate cyclase. This is far easier to demonstrate in the intermediate lobe, where all cells appear to respond to dopamine, than in the anterior lobe, where the dopamine-sensitive cells probably represent only a small minority.

In terms of multiple receptor categories, the suggestion made by Kebabian and Calne (2) of a distinction between D-1 and D-2 subtypes, based on whether the receptors lead to stimulation of adenylate cyclase or not, has been widely accepted. It now seems that in many cases (perhaps all) the D-2 sites are also coupled to adenylate cyclase, although in an inhibitory rather than stimulatory manner. All of the studies on pituitary place the dopamine receptors there clearly in the D-2 category. M. Caron et al., however, report interesting new results that indicate that these sites can exist in more than one form—with about half of the sites in the resting state in a form with high affinity for agonists and half in a low agonist affinity state. These forms can be interconverted, and guanyl nucleotide or NEM treatment shifts the population mainly to the agonist high affinity form. This may also help to explain the observations of Kebabian et al. that agonists were far more potent in intact pituitary cell preparations than in broken cell preparations.

There remain many difficulties in further understanding the nature of the different dopamine receptor categories. We continue to lack suitably selective agonists or antagonists for the D-1 and D-2 sites. In terms of agonists, bromocriptine and related ergoline derivatives are still the most selective D-2-stimulants, and the series of benzazepines related to SK&F 38393 are the most promising D-1-selective agents (J. Weinstock et al.). The discovery that benzazepines act in a stereochemically specific manner, and the resolution of the active and inactive stereoisomeric forms, offers further hope for more selective agonists in future. D. E. Nichols provides a detailed and thoughtful review of the medicinal chemistry aspects of dopamine agonist design. There is still no D-1-selective antagonist, although sulpiride and related benzamides are widely used as selective D-2-antagonists. The availability of at least one tissue model for D-1 receptors, the stimulatory effects of dopamine on parathyroid hormone secretion from bovine parathyroid cells, is of considerable importance, but we still have no corresponding model to elucidate the possible function of D-1 sites in the CNS.

There is also considerable difficulty in relating the biochemical classification of D-1 and D-2 sites to the pharmacologically defined dopamine receptor subtypes, described by L. Goldberg and J D. Kohli, on the basis of their painstaking analysis of agonist/antagonist actions on cardiovascular responses. They describe two subcategories of dopamine receptors, but these *do not* correspond readily to D-1 and D-2 sites. Thus, their "DA$_1$" receptors that cause relaxation of vascular smooth muscle have an agonist specificity similar to the D-1 sites: with an absolute requirement for a catechol grouping; rigid catechol analogues such as ADTN are fully active; ergolines are inactive. The responses are blocked by neuroleptics, but unlike the D-1 site, which is quite unresponsive to sulpiride, the DA$_1$ receptors are potently blocked by sulpiride. The "DA$_2$" receptors, mediating presynaptic control of norepinephrine release from sympathetic nerve terminals, resemble D-2 sites in their specificity, but again the comparison is not precise. Interestingly, the DA$_2$ sites can be stimulated even by some monophenolic agonists. This is intriguing because some of the newly described "autoreceptor agonists" in CNS, such as 3-PPP (*3*) are monophenolic structures. The topical question of whether the receptors located on the surface of dopamine neurons in CNS (autoreceptors) represent a unique pharmacological class does not perhaps receive as much attention in this volume as it should have, although the issue is discussed in some detail by J. C. Stoof. Although such receptors clearly resemble the D-2 class in many respects, there remains the suspicion that there may be some subtle differences.

Despite the large effort already directed to studies of dopamine and dopamine receptors it is clear that many questions remain unanswered. The area continues to be one of considerable promise and intellectual vigor and from this ferment useful new pharmacological and, possibly, new therapeutic tools may eventually emerge.

Literature Cited

1. Seeman, P. *Pharmac. Rev.* **1980**, *32*, 229–313.
2. Kebabian, J. W.; Calne, D. B. *Nature (London)* **1979**, *277*, 93–96.
3. Hjorth, S.; Carlsson, A.; Wikström, H.; Lindberg, P.; Sanchez, D.; Hacksell, U.; Arvidsson, L. E.; Svensson, U.; Nilsson, J. L. G. *Life Sci.* **1981**, *28*, 1225–1238.

LESLIE L. IVERSEN
MRC Neurochemical Pharmacology Unit
Medical Research Council Centre
Hills Road
Cambridge CB2 2QH, United Kingdom

D-1 Dopamine Receptor-Mediated Activation of Adenylate Cyclase, cAMP Accumulation, and PTH Release in Dispersed Bovine Parathyroid Cells

EDWARD M. BROWN and BESS DAWSON-HUGHES

Brigham and Women's Hospital, Endocrine-Hypertension Unit, Boston, MA 02115

The evidence supporting the existence of a specific category of dopamine receptor on the parenchymal cells of the bovine parathyroid gland and the possible biochemical mechanisms by which dopamine stimulates the release of parathyroid hormone are reviewed. The dopamine receptor on the bovine parathyroid cell is compared to other dopamine receptors.

The parathyroid glands play a major role in normal calcium homeostasis (1). Calcium is generally recognized as the principal physiological regulator of the release of parathyroid hormone (PTH) (2). When the plasma concentration of ionized calcium decreases, PTH secretion increases. In turn, this PTH acts to raise plasma calcium by three mechanism: first, PTH increases renal tubular reabsorption of calcium; second, PTH enhances the release of skeletal calcium; and third, PTH increases gastrointestinal absorption of calcium by stimulating renal formation of 1,25 dihydroxyvitamin D. These three mechanisms elevate the plasma concentration of ionized calcium and consequently reduce the augmented secretion of PTH, thereby closing a negative feedback loop. The inhibitory effect of calcium upon PTH secretion contrasts with the stimulatory effect of calcium upon most other secretory systems (3). In addition to calcium, other factors also modify PTH secretion. Many of these factors change cellular cyclic adenosine 3'5', monophosphate (cAMP) levels at the same time that they modify PTH secretion (4,5,6).

This volume provides a forum in which it is appropriate to discuss the bovine parathyroid gland and the effects of dopamine upon this tissue. When administered intravenously to cows, dopamine raises the plasma content of immunoreactive PTH (7). This stimulatory effect of dopamine is partially blocked by pimozide, a dopamine antagonist, but is unaffected by propranolol, a beta-adrenergic antagonist. An understanding of

0097–6156/83/0224–0001$06.25/0

the cellular mechanisms involved in mediating this
dopamine-induced stimulation of PTH secretion is limited by the
cellular heterogeneity of the bovine parathyroid gland (7) as
well as by the temporal and spatial imprecision of intravenous
infusions. Nevertheless, the in vivo results provide a
standard against which in vitro results can be compared.

The cellular and molecular events involved in the
dopamine-stimulated release of PTH can be clarified in
experiments utilizing bovine parathyroid cells dispersed with
collagenase and DNase (8). This dispersion procedure yields
parenchymal cells with only a slight contamination by red blood
cells. The parenchymal cells exclude trypan blue and appear
normal by light and electron microscopy (8). These cells
release PTH in a linear fashion for several hours; the release
is inhibited by calcium and stimulated by dopamine and
beta-adrenergic agonists at concentrations comparable to those
used to elicit physiological responses in vivo (4,8).

Dopamine Enhances PTH Secretion in Dispersed Bovine Parathyroid Cells

Dopamine (1 µM) causes a transient 2 to 4-fold increase in
the rate of release of immunoreactive PTH (IR-PTH) from
dispersed bovine parathyroid cells (9). The stimulatory effect
of dopamine is maximal after 5 minutes exposure and persists
for approximately 30 minutes (Figure 1). Several compounds
mimicking the effects of dopamine in other systems mimic the
stimulatory effect of dopamine on IR-PTH release. Both
2-amino, 6,7-dihydroxy tetralin (6,7-ADTN) and SKF 38393
increase the release of IR-PTH to the same degree as does
dopamine. The release of IR-PTH is half-maximally stimulated
by dopamine, 6,7-ADTN and SKF 38393 at 0.2 µM, 0.15 µM and
0.3 µM, respectively (Figure 2) (10). In contrast, other
dopaminergic agonists are substantially less potent than
dopamine; both apomorphine and lergotrile elicit no more than
25% of the maximal response to dopamine, and lisuride is devoid
of agonist activity. The dopamine-stimulated release of IR-PTH
is inhibited in a stereospecific manner by the isomers of
flupenthixol (9); cis-flupenthixol is approximately 100-fold
more potent than its trans-isomer (calculated K_i's of 33 nM and
3,300 nM, respectively) (Figure 3).

Dopamine Enhances cAMP Accumulation in Dispersed Bovine Parathyroid Cells

Dopamine causes a 20 to 30-fold increase in the content of
cAMP in dispersed bovine parathyroid cells (Figure 4) (9).
Like the dopamine-stimulated enhancement of PTH release, the
dopamine-stimulated increase in cAMP content is maximal after 5
to 10 minutes of exposure to 10 µM dopamine (9) and

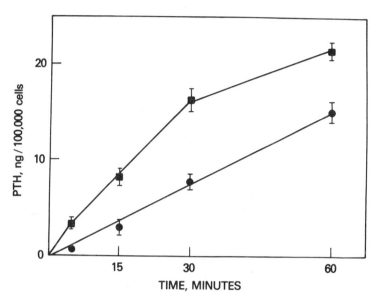

Figure 1. Stimulation of PTH release from dispersed bovine parathyroid cells by dopamine. Cells were incubated with (■) or without (●) 1 μM dopamine, and PTH release was determined by radioimmunoassay.

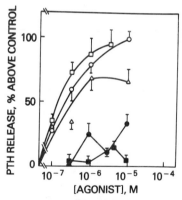

Figure 2. Stimulation of PTH release from dispersed bovine parathyroid cells by varying concentrations of dopamine (○), 6,7-ADTN (□), SKF 38393 (△), apomorphine (●), or dihydroergocryptine (■). (Reproduced with permission from Ref. 10. Copyright 1980, American Society for Pharmacology and Experimental Therapeutics.)

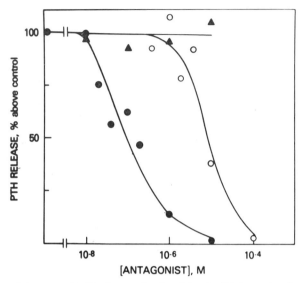

Figure 3.   Inhibition of PTH-release stimulated by 1 μM dopamine by α-flupen-thixol (●), β-flupenthixol (○), or (−) propranolol (▲).

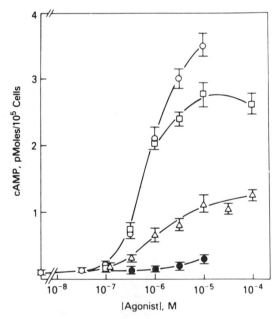

Figure 4.   Stimulation of cAMP accumulation in dispersed bovine parathyroid cells by varying concentrations of dopamine (○), 6,7-ADTN (□), SKF 38393 (△), or apomorphine (●). (Reproduced with permission from Ref. 10. Copyright 1980, American Society for Pharmacology and Experimental Therapeutics.)

subsequently decreases (Figure 5). Significant quantities of
cAMP are excreted from the dispersed cells during the first 15
minutes of exposure to dopamine. As is the case for the
dopamine-stimulated enhancement of IR-PTH release, several
dopaminergic agonists mimic the ability of dopamine to increase
cAMP accumulation in the dispersed bovine parathyroid cells
(Table I). For each compound tested, the concentration of
agonist half-maximally enhancing cAMP accumulation is
approximately the same as the concentration of agonist required
to half-maximally stimulate the release of IR-PTH. Dopamine
antagonists block the dopamine-stimulated increase in cAMP
(Figure 6) (Table I). Interestingly, apomorphine, lisuride,
lergotrile and bromocriptine each block the dopamine receptor
in this system (Figure 6).

Dopamine Stimulates Adenylate Cyclase Activity of Dispersed Bovine Parathyroid Cells

When tested on osmotically-lysed bovine parathyroid cells,
dopamine enhances the activity of adenylate cyclase, the enzyme
converting ATP to cAMP (11). In comparison with its effect on
cAMP accumulation, the effect of dopamine on adenylate cyclase
activity is relatively modest, only a 2-fold increase in enzyme
activity (Figure 7). Guanosine 5'-triphosphate (GTP) increases
the stimulatory effect of dopamine; in the presence of GTP,
there is 3 to 4-fold stimulation of enzyme activity (11).
    In other cell types, guanine nucleotides interact with a
guanine nucleotide subunit (G- or N_s-subunit) to translate
receptor stimulation into increased adenylate cyclase activity
(12). Cholera toxin inhibits a specific GTPase on this guanine
nucleotide subunit and thereby increases adenylate cyclase
activity (13). In dispersed cells from the bovine parathyroid
gland, cholera toxin markedly increases cAMP formation and
causes a 3 to 10-fold increase in the apparent affinity of
dopamine for its receptor (as determined by cAMP accumulation
or IR-PTH secretion (14). The effects of guanine nucleotides
and cholera toxin on cAMP accumulation in parathyroid cells
result from interactions with the guanine nucleotide subunit in
this cell.
    In either the presence or absence of GTP, half-maximal
stimulation of enzyme activity is achieved with 3 μM dopamine.
Both 6,7-ADTN and epinine (N-methyl dopamine) stimulate
adenylate cyclase activity to the same degree as does dopamine
(Figure 8). In contrast, apomorphine is a partial agonist
eliciting only 30% of the maximal effect of dopamine. The
dopamine-stimulated adenylate cyclase activity is selectively
blocked by cis-flupenthixol rather than the trans-isomer of
this antagonist (11). Among the antagonists tested, the order
of potency is: cis-flupenthixol = fluphenazine >
chlorpromazine > haloperidol > trans-flupenthixol (Table I).

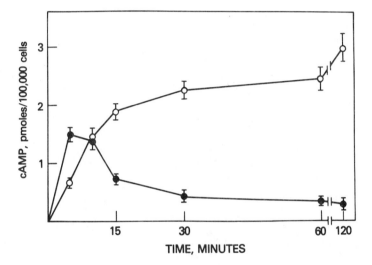

Figure 5.   Stimulation of intracellular (●) and extracellular (○) cAMP by 10 μM dopamine in dispersed bovine parathyroid cells. cAMP was determined by radio-immunoassay.

TABLE I

Affinity of drugs for the dopamine receptor in bovine parathyroid cells determined in experiments measuring cAMP
accumulation, adenylate cyclase, or PTH release.

$$K_a \text{ or } K_i \text{ } (\mu M)$$

	cAMP Accumulation	Adenylate Cyclase	PTH Release
Agonists			
Dopamine	0.6	3	0.2
ADTN	0.5	4	0.15
Epinine	0.6	10	--
Partial Agonists			
SKF 38393	1	3	0.3
Apomorphine	1	3	1
Antagonists			
d-Butaclamol	.005	--	--
α-Flupenthixol	.03	.011	.033
Fluphenazine	.04	.07	--
Lisuride	.015	.16	.09
Lergotrile	.7	.37	.9
Bromoergocryptine	21	2.1	--
YM-09151-2	15	--	--
(+) Sulpiride	13	--	--
(-) Sulpiride	$^{1\!/}$	--	--

$1\!/$ 20% inhibition at 3 x 10^{-5}

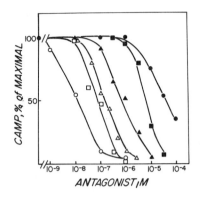

Figure 6. Inhibition of cAMP accumulation stimulated by 1 μM dopamine by lisuride (○), α-flupenthixol (□), β-flupenthixol (■), lergotrile (▲), α-bromoergocryptine (●), or fluphenazine (△).

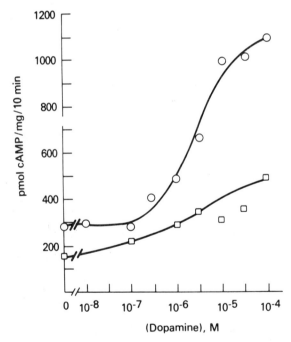

Figure 7. Dopamine-stimulated adenylate cyclase activity in lysates of bovine parathyroid cells in the absence (□) or presence (○) of 100 μM guanosine triphosphate (GTP). (Reproduced with permission from Ref. 11. Copyright 1980, The Endocrine Society.)

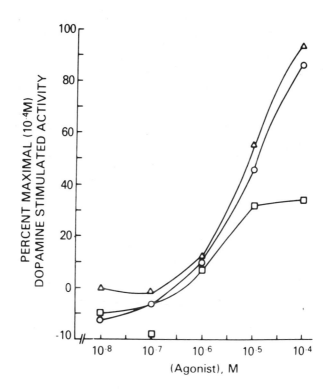

Figure 8.   Stimulation of adenylate cyclase in lysates of bovine parathyroid cells by 6,7-ADTN (△), epinine (○), or apomorphine (□) in the presence of 100 μM GTP. (Reproduced with permission from Ref. 11. Copyright 1980, The Endocrine Society.)

Does Dopamine Interact with Beta- or Alpha-adrenergic receptors?

The bovine parathyroid gland possesses a beta-adrenergic receptor. Like dopamine, beta-adrenergic agonists enhance adenylate cyclase activity, cAMP accumulation, and the release of IR-PTH (15). The beta-adrenergic receptor, however, can be differentiated from the receptor for dopamine with selective antagonists (Figure 9). For example, propranolol, a potent antagonist of the beta-adrenergic receptor, causes a nearly complete inhibition of cAMP accumulation stimulated by isoproterenol, epinephrine, or norepinephrine. Propranolol, on the other hand, does not diminish dopamine-stimulated cAMP accumulation. Conversely, cis-flupenthixol blocks the dopamine- and epinine-stimulated accumulation of cAMP but does not reduce the isoproterenol-stimulated accumulation of cAMP.

The bovine parathyroid gland also possesses an alpha-adrenergic receptor (16). However, it seems improbable that an interaction between dopamine and the alpha-adrenergic receptor accounts for the physiological and biochemical effects of dopamine upon this tissue. Phentolamine, an alpha-adrenergic antagonist, has no effect on dopamine-stimulated cAMP accumulation at concentrations as high as 10 μM, a concentration totally blocking alpha-adrenergic effects in this system (10). Furthermore, unlike the dopaminergic receptor, stimulation of the parathyroid alpha-adrenergic receptor inhibits the agonist-stimulated activation of accumulation of cAMP and release of IR-PTH (16).

Several dopaminergic drugs interact not only with the dopamine receptor but also with alpha- and beta-adrenergic receptors in the bovine parathyroid gland. For example, lisuride, a potent dopamine agonist upon the anterior pituitary gland, blocks the alpha-adrenergic receptor, the beta-adrenergic receptor, and the receptor for dopamine in parathyroid cells (11).

Does cAMP Trigger the Release of IR-PTH?

The dopamine-stimulated formation of cAMP may initiate the dopamine-induced release of IR-PTH. A linear relationship exists between the dopamine-induced release of IR-PTH and the logarithm of the dopamine-induced accumulation of cAMP (17). Similarly, other agents increasing cAMP accumulation and IR-PTH release (e.g. beta-adrenergic agonists, secretin and phosphodiesterase inhibitors, also display such a log-linear relationship. Additional support for the possibility that intracellular cAMP might initiate PTH secretion comes from the observations that cholera toxin (14), phosphodiesterase inhibitors (17) and dibutyryl cAMP (18), agents known to increase intracellular cAMP or mimic the biochemical effects of cAMP, increase the release of IR-PTH.

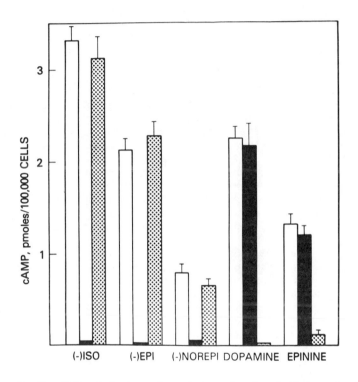

Figure 9.   Specificity of β-adrenergic and dopaminergic stimulation of cAMP accumulation.   Dispersed parathyroid cells were incubated with the indicated β-adrenergic or dopaminergic agonists either alone (open bars), with 1 μM (−) propranolol (solid bars), or with 10 μM α-flupenthixol (stippled bars).

A cAMP-dependent phosphorylation of specific cellular substrates is hypothesized to initiate tissue-specific physiological responses. Certain elements of this hypothesis can be applied to the bovine parathyroid gland. Dispersed bovine parathyroid cells contain predominantly the type 2 isozyme(s) of cAMP-dependent protein kinase (19). The activation state of this form of the kinase remains relatively constant when tissue is extracted into buffers containing high concentrations of salt. This situation permits an estimation of the drug-induced activation of the cAMP-dependent protein kinase activity in intact cells. In a recent series of experiments utilizing dispersed bovine parathyroid cells, we compared the dopamine-induced release of IR-PTH with dopamine-induced alterations in the activity ratio of the cAMP-dependent protein kinase (an estimate of the fractional activation of this enzyme) (20). A close correlation exists between dopamine-induced changes in the activity ratio and dopamine-induced IR-PTH secretion (Figure 10). This observation is consistent with a mediatory role for cAMP-dependent protein phosphorylation in dopamine-stimulated hormonal secretion. Additional evidence supporting this possibility is the observation that cAMP promotes the phosphorylation of several endogenous proteins in sonicates of dispersed bovine parathyroid cells (21) and that dopamine stimulates the phosphorylation of two proteins of molecular weight 15,000 and 19,000 in intact bovine parathyroid cells (22).

The biochemical mechanisms by which cAMP-dependent phosphorylation leads to enhanced IR-PTH release remain to be determined. It is of interest, however, that isoproterenol activates phosphorylation of proteins of similar molecular weight in the rat parotid gland (23), while glucagon stimulates phosphorylation of a protein of molecular weight 19,000 in calcitonin-secreting cultured cells from a medullary carcinoma of the rat thyroid (24). It is conceivable that in all three tissues, activation of exocytosis results from a cAMP-dependent phosphorylation of a critical cellular substrate.

Receptors Inhibiting Dopamine-stimulated cAMP Accumulation and PTH Secretion

Both alpha-adrenergic agonists (16) and prostaglandin F2α (PGF2α) (25) diminish dopamine-stimulated cAMP accumulation and hormonal secretion. The inhibitory effects of these agents on hormonal secretion can be quantitatively accounted for by their inhibitory effect on cAMP accumulation (17). The mechanism(s) by which these compounds lower cellular cAMP has not been investigated; in other systems alpha-adrenergic agonists inhibit adenylate cyclase through a GTP-dependent mechanism

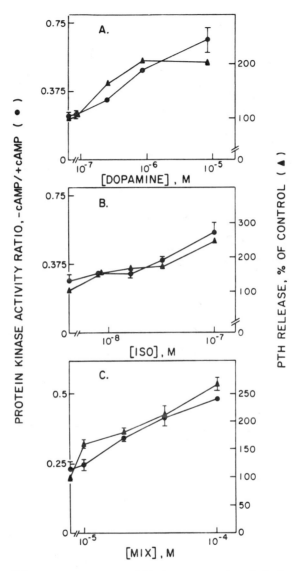

Figure 10.   Effects of isoproterenol (ISO), dopamine, and methylisobutylxanthine (MIX) on protein kinase activation and PTH secretion.

Dispersed cells were incubated with increasing concentrations of the indicated secretagogue. The cells were sedimented, PTH was determined in the supernatant, and protein kinase activity was determined with and without cAMP in sonicates of the cellular pellet. Protein kinase activities are expressed as activity ratios [(activity − cAMP)/(activity + cAMP)] which is equivalent to the fractional activation of the enzyme. (Reproduced with permission from Ref. 20. Copyright 1982, The Endocrine Society.)

(26), which may be analogous to that utilized by the pituitary
D-2 receptor (see below and Kebabian et al. this volume).

Interaction of Dopamine and Calcium in Regulating PTH Release in Dispersed Bovine Parathyroid Cells

Calcium lowers cellular cAMP and inhibits both basal and
agonist-stimulated IR-PTH secretion (17). The ability of
calcium to lower cAMP cannot quantitatively account for the
inhibition of hormone release, however, suggesting that calcium
exerts inhibitory effects independent of its actions of cAMP
(17). This possibility is further supported by the finding
that calcium inhibits agonist-stimulated PTH release without
altering the enhancement of cAMP-dependent protein kinase
activity (20). Similarly, increased extracellular calcium does
not affect agonist-stimulated phosphorylation of parathyroid
proteins (22). Since the divalent cation ionophore A23187
reduces dopamine-stimulated secretion in a fashion similar to
that of elevated extracellular calcium, it is possible that
changes in the cytosolic calcium concentration are important in
regulating the effects of extracellular calcium on PTH release
(27).

Regulation of PTH Release Through cAMP-dependent and cAMP-independent Pathways

A schematic representation of the cellular mechanisms
through which agonists, such as dopamine, stimulate PTH release
and calcium inhibits hormonal secretion is shown in Figure 11.
In this schema, cAMP is a stimulatory second messenger while
cytosolic calcium serves to inhibit hormone release by acting
at several loci within the cell. The detailed molecular
mechanisms through which cAMP and cellular calcium modulate
cellular function remain to be determined.

Possible Role of Dopamine in Regulating PTH Secretion In Vivo

The physiological role of dopamine and other
catecholamines in PTH secretion is unknown. It is of interest
that large quantities of dopamine (3.9 to 13.9 pg/g) occur in
the bovine parathyroid gland. This dopamine is localized
within mast cells (28) which occur throughout the gland (7).
In contrast, the content of norepinephrine is substantially
lower than the content of dopamine (28).
Catecholamine-containing neurons do not innervate the
parenchymal cells of the bovine parathyroid gland; only an
occasional norepinephrine-containing neuron terminating upon a
blood vessel is demonstrated by fluorescence histochemistry
(27).
Because of the large content of dopamine in the mast cells

within the bovine parathyroid gland, this catecholamine may play some role in the regulation of PTH release in vivo. To date, there is no experimental evidence of any such role for dopamine in this tissue. Unlike bovine mast cells, human mast cells contain histamine rather than dopamine. Interestingly, human parathyroid glands respond to histamine but not dopamine (29).

Effects of Dopaminergic Antagonists on Calmodulin-stimulated Phosphodiesterase Activity

Both calmodulin and a calcium- and calmodulin-dependent phosphodiesterase activity occur in dispersed bovine parathyroid cells (30). Some of the phenothiazines blocking the dopamine receptor in this tissue inhibit the interaction between calmodulin and the calmodulin-dependent phosphodiesterase. A series of experiments have tested the possibility that calmodulin might play a role in the effects of these agents on the parathyroid gland. The dopamine antagonists examined were weak inhibitors of calmodulin-stimulated phosphodiesterase activity (Table II). The stereospecific dopaminergic antagonists (-) and (+) butaclamol and cis- and trans-flupenthixol were of equal potency in their effects on calmodulin-dependent phosphodiesterase activity. This contrasts with the marked difference in the potency of each pair of stereoisomers as dopamine antagonists. It appears unlikely that the calmodulin-dependent phosphodiesterase activity (and by inference calmodulin) plays any role in the actions of dopaminergic antagonists upon the bovine parathyroid gland. The effects of calmodulin upon adenylate cyclase activity of the bovine parathyroid gland have not been determined. However, EGTA does not diminish adenylate cyclase activity; this suggests that calcium (and by inference calmodulin) may not be an important cofactor for this enzyme in the bovine parathyroid gland (E.M. Brown and B. Dawson-Hughes, unpublished observations).

Relationship of Dopamine Receptors on Bovine Parathyroid Cells to Other Dopamine Receptors: Dopamine Receptors Stimulating Adenylate Cyclase Activity

In order to account for the physiological and biochemical effects of dopamine on the bovine parathyroid gland, it is necessary to postulate the existence of a receptor for dopamine upon the surface of the parenchymal cells. This receptor is characterized by dopamine, epinine, and 6,7-ADTN being full agonists, apomorphine and SKF 38393 being partial agonists, and ergots, such as lisuride and lergotrile, being devoid of agonist properties (Table I). Among agonists, (+) butaclamol,

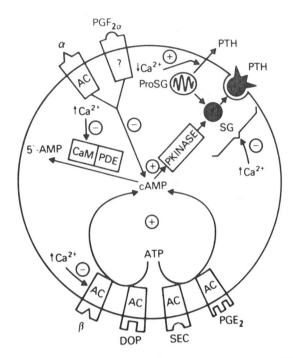

Figure 11.   Schematic diagram of the regulation by calcium and cAMP of PTH release.

B, DOP, SEC, and PGE₂ indicate the receptors for β-adrenergic catecholamines, dopaminergic catecholamines, secretin, and prostaglandin E_2, respectively, which stimulate cAMP accumulation through activation of adenylate cyclase (AC). Receptors for α-adrenergic catecholamines α and prostaglandin $F_2α$ PGF₂α reduce cAMP content. cAMP is degraded by phosphodiesterase (PDE), one form of which is activated by calmodulin (CaM). The effects of cAMP are presumably mediated by cAMP-dependent protein kinase (PKINASE), which leads to exocytosis activation of PTH-containing secretory granules (SG). Possible sites of calcium action are shown: inhibiting adenylate cyclase, activating phosphodiesterase, inhibiting PTH secretion distal to cAMP formation, or stimulating (low calcium) the release of immature secretory granules (ProSG). Key: +, positive effect on secretion of PTH; and —, inhibitory effect on secretion. *(Reproduced with permission from Ref. 29. Copyright 1980, Academic Press.)*

TABLE II

Effects of dopaminergic antagonists on dopamine-stimulated cAMP accumulation, adenylate cyclase activity, and calmodulin-stimulated phosphodiesterase (PDE) activity in intact bovine parathyroid cells or cellular homogenates. Values for K_i or IC_{50} are given as μM. NT, not tested.

Antagonist	K_i for cAMP	K_i for Cyclase	IC_{50} for PDE
Fluphenazine	.07	.04	15
α–Flupenthixol	.060	.030	32
β–Flupenthixol	1.3	†	24
d–Butaclamol	.0045	NT	>100
ℓ–Butaclamol	†	NT	>100

† no inhibition at 10^{-4} M.

cis-flupenthixol, and lisuride are most potent.
Phenothiazines, such as fluphenazine and chlorpromazine, have
K_i's in the range of 4×10^{-8}–4×10^{-7} M. Most ergots (e.g.
lergotrile, dihydroergocryptine, and bromocryptine) are of
moderate to low potency ($\sim$ 1-10 μM). (+) and (-) Butaclamol
and cis- and trans-flupenthixol show marked stereospecificity,
with the active isomer being of high potency. Agents
putatively specific for dopamine receptors in the anterior
pituitary or intermediate lobe of the rat pituitary are either
weak antagonists [(+)- and (-)-sulpiride, YM-09151-2 (31)] or
have no effect [LY-141865 (32)] in bovine parathyroid cells.

Several features of the dopamine receptor in the bovine
parathyroid gland also occur in other mammalian tissues as well
as in lower vertebrates and even invertebrates (Table III).
These dopamine-sensitive adenylate cyclase systems are
characterized by dopamine, epinine and 6,7-ADTN being full
agonists of micromolar potency, apomorphine being a partial
agonist, and ergots such as lisuride, lergotrile, or
bromocriptine displaying minimal agonist activity but being
antagonists of moderate potency. In the preparations of intact
cells which have been tested, dopamine agonists increase cAMP
accumulation. This dopamine-stimulated increase in cAMP can be
linked to activation of a cAMP-dependent protein kinase (37),
or to phosphorylation of specific cellular substrates (41). In
some cases, exogenous cAMP mimics the physiological effects of
dopamine (42). Thus, these dopamine receptors (designated as
D-1 receptors, (43) appear to act, at least in part, through
elevation of cellular cAMP, activation of cAMP-dependent
protein kinases, and phosphorylation of specific cellular
substrates.

Dopamine Receptors Inhibiting Adenylate Cyclase Activity

Certain dopamine receptors do not resemble the dopamine
receptor in the bovine parathyroid gland. For example, the
dopamine receptor on the mammotroph of the pituitary gland is
characterized by dopamine, apomorphine, and ergot alkaloids
being agonists with nanomolar potency. Butyrophenones are
potent antagonists of this receptor. While this class of
receptor was originally classified on the basis of the lack of
any stimulatory effects on cAMP accumulation or adenylate
cyclase activity (43), it has become clear that stimulation of
this receptor inhibits cAMP metabolism (44). An additional
example of this second category of receptor (designated as the
D-2 receptor) occurs in the intermediate lobe of the rat
pituitary gland. In this latter system, dopamine inhibits
basal and particularly agonist-stimulated cAMP accumulation and
adenylate cyclase activity (45,46); these inhibitory effects
are GTP-dependent (47). The changes in cellular cAMP correlate
with the inhibition of hormone release from the IL (45,46,
Kebabian et al., this volume).

TABLE III

Ka for dopamine and Ki for several dopamine antagonists on dopamine-sensitive adenylate cyclase

Tissue	K_a, μM Dopamine	α-Flupenthixol	K_i, nM Fluphenazine	Spiroperidol	Reference
Bovine Parathyroid Cell	3	11	40	3,000	9,10
Rat Candate Nucleus	10	1	4-8	95	33
Substantia Nigra	2-10	0.48	67[1]	5.3[2]	34
Rat Olfactory Tubercle	10	----	5	200[2]	35
Superior Cervical Ganglion	6-10	----	---	----	36
Guinea Pig Retina	10	4.8	3.9	----	37
Carp Retina	1-2	5	3	600	38
Snail Nervous System	2	----	~50	----	39
Cockroach Brain	1.7	----	20	----	40

[1] K_i for chlorpromazine. Fluphenazine is about 7-10-fold more potent than chlorpromazine in several dopamine-sensitive adenylate cyclase systems.

[2] K_i for haloperidol. Spiroperidol and haloperidol are of comparable potency in several dopamine-sensitive adenylate cyclase systems.

There are distinct parallels, therefore, between the D-1 and D-2 receptors. Both interact with adenylate cyclase, one through a stimulatory guanine nucleotide subunit and the other, presumably, through an inhibitory subunit (47). Both may exert their biological effects through changes in cellular cAMP (as was initially hypothesized by Sutherland and his colleagues). The relationship between the D-1 and the D-2 receptors is very analogous to that between the beta- and the α_2-adrenergic receptors. These receptors also act upon adenylate cyclase through stimulatory and inhibitory guanine nucleotide subunits, respectively (12,48). The α_1-adrenergic receptor, on the other hand, is thought to act through changes in cellular calcium dynamics, without appreciable effects on adenylate cyclase (49). By analogy, it may be speculated that an additional class of dopamine receptor acting through changes in cytosolic calcium (and, therefore, equivalent to the α_1- adrenergic receptor) might exist. Conceivably, this hypothesized category of dopamine receptor would correspond to the D_o receptor postulated to exist by Meunier and Labrie (50). At present, no example of either of these theoretical constructs has been identified.

Summary

Dispersed bovine parathyroid cells contain a dopamine receptor which increases cellular cAMP through a guanine nucleotide-stimulated activation of adenylate cyclase. The dopamine-stimulated increase in cellular cAMP correlates closely with activation of cAMP-dependent protein kinase, phosphorylation of endogenous cellular proteins, and secretion of IR-PTH. The potency of various dopaminergic agonists and antagonists in modifying these biological effects as well as the role of cAMP in modifying physiological function suggest that the bovine parathyroid dopamine receptor is a D-1 dopamine receptor. This system may be of use in studying the interaction of radiolabeled ligands with the D-1 receptor.

Acknowledgments

The authors gratefully acknowledge the excellent technical help of Joseph Thatcher and Edward Watson and secretarial work of Mrs. Nancy Orgill. This work was supported by USPHS Grants AM25910 and AM30028.

Literature Cited

1. Parsons, J.A. "Endocrinology"; Grune and Stratton: New York, 1979; Vol 2, p 621.
2. Sherwood, L.M.; Potts, Jr., J.T.; Care, A.D.; Mayer, G.P.; Aurbach, G.D. Nature 1966, 209, 52.

3. Douglas, W.W. Ciba Foundation Symposium 1978, 54, 61.
4. Brown, E.M. Mineral Electrolyte Metabolism 1982, 130, 3.
5. Peck, W.A.; Klahr, S. "Advances in Cyclic Nucleotides Research"; Raven Press: New York, 1979; Vol 11, p 89.
6. Heath, III, H. Endocrine Reviews 1980, 1, 319.
7. Blum, J.W.; Kunz, P.; Fischer, J.A.; Binswanger, U.; Lichtensteiger, W.; DaPrada, M. Am. J. Physiol. 1980, 239, E255.
8. Brown, E.M.; Hurwitz, S.; Aurbach, G.D. Endocrinology 1976, 99, 1582.
9. Brown, E.M.; Carroll, R.; Aurbach, G.D. Proc. Nat'l. Acad. Sci. USA 1977, 74, 4210.
10. Brown, E.M.; Attie, M.F.; Reen, S.; Gardner, D.G.; Kebabian, J.; Aurbach, G.D. Molec. Pharmacol 1980, 18, 335.
11. Attie, M.F.; Brown, E.M.; Gardner, D.G.; Spiegel, A.M.; Aurbach, G.D. Endocrinology 1980, 107, 1776.
12. Rodbell, M. Nature 1980, 284, 17.
13. Cassel, D.; Selinger, Z. Proc. Nat'l. Acad. Sci. USA 1977, 74, 3307.
14. Brown, E.M.; Gardner, D.G.; Windeck, R.A.; Aurbach, G.D. Endocrinology 1979, 104, 218.
15. Brown, E.M.; Hurwitz, S.; Aurbach, G.D. Endocrinology 1977, 100, 1696.
16. Brown, E.M.; Hurwitz, S.H.; Aurbach, G.D. Endocrinology 1978, 103, 893.
17. Brown, E.M.; Gardner, D.G.; Windeck, R.A.; Aurbach, G.D. Endocrinology 1978, 103, 2323.
18. Morrissey, J.J.; Cohn, D.V. J. Cell Biol. 1979, 83, 521.
19. Thatcher, J.G.; Gardner, D.G.; Brown, E.M. Endocrinology 1982, 110, 1367.
20. Brown, E.M.; Thatcher, J.G. Endocrinology 1982, 110, 1374.
21. Brown, E.M.; Thatcher, J.G. Program and Abstracts. Fourth Annual Scientific Meeting of the American Society for Bone and Mineral Research, San Francisco, CA, 1982, p S-37.
22. Lasker, R.D.; Spiegel, A.M. Clin. Res. 1982, 30, 398A.
23. Baum, B.J.; Freiberg, J.M.; Ito, H.; Roth, G.S.; Filburn, C.R. J. Biol. Chem. 1981, 256, 9731.
24. Gagel, R.F.; Andrews, K.L. Abstracts of the 4th Annual Scientific Meeting of the American Society for Bone and Mineral Research 1982, p S-18.
25. Gardner, D.G.; Brown, E.M.; Windeck, R.; Aurbach, G.D. Endocrinology 1979, 104, 1.
26. Jakobs, K.H.; Saur, W.; Schulz, G. FEBS Letters 1978, 85, 167.
27. Brown, E.M.; Gardner, D.G.; Aurbach, G.D. Endocrinology 1980, 106, 133.
28. Jacobowitz, D.; Brown, E.M. Experentia 1980, 36, 115.
29. Brown, E.M.; Aurbach, G.D. Vitamins and Hormones 1980, 38, 205.

30. Brown, E.M. Endocrinology 1980, 107, 1998.
31. Grewe, G.W.; Frey, E.A.; Cote, T.E.; Kebabian, J.W. Eur. J. Pharmacol. 1982, 81, 149.
32. Tsuruta, K.; Frey, E.A.; Grewe, C.W.; Cote, T.E.; Eskay, R.L.; Kebabian, J.W. Nature, 1981, 292, 463.
33. Kebabian, J.W.; Petzgold, G.L.; Greengard, P. Proc. Nat'l. Acad. Sci. 1972, 69, 2145.
34. Phillipson, O.T.; Horn, A.S. Nature 1976, 261, 418.
35. Horn, A.S.; Cuello, A.C.; Miller, R.J. J. Neurochem. 1974, 22, 265.
36. Kebabian, J.W.; Greengard, P. Science 1971, 174, 1346.
37. Brown, J.H.; Makman, M.H. Proc. Nat'l. Acad. Sci. USA 1972, 69, 539.
38. Watling, K.J.; Dowling, J.E. J. Neurochem. 1981, 36, 559.
39. Osborne, N.N. Experentia 1977, 33, 917.
40. Harmar, A.J.; Horn, A.S. Mol. Pharmacol. 1977, 13, 512.
41. Nestler, E.J.; Greengard, P. Proc. Nat'l. Acad. Sci. USA 1980, 77, 7479.
42. Greengard, P.; McAfee, D.A.; Kebabian, J.W. Adv. Cyclic Nucl. Res. 1972, 1, 373.
43. Kebabian, J.W.; Calne, D.B. Nature 1979, 277, 93.
44. Camilli, P.D.; Macconi, D.; Spada, A. Nature (London) 1979, 278, 252.
45. Munemura, M.; Eskay, R.L.; Kebabian, J.W. Endocrinology 1980, 106, 1795.
46. Cote, T.E.; Grewe, C.W.; Kebabian, J.W. Endocrinology 1981, 108, 420.
47. Cote, T.E.; Grewe, C.W., Tsuruta, K.; Stoof, J.C.; Eskay, R.L.; Kebabian, J.W. Endocrinology 1982, 110, 812.
48. Brown, E.M.; Aurbach, G.D. "Contemporary Metabolism"; Plenum: New York, 1982; Vol 2, p 247.
49. Exton, J.H. Am. J. Physiol. 1980, Vol., E3.
50. Meunier, H.; Labrie, F. Life Science 1982, 30, 963.

RECEIVED March 25, 1983

Commentary: Dopamine-Sensitive Adenylate Cyclase as a Receptor Site

PIERRE M. LADURON

Janssen Pharmaceutica, Department of Biochemical Pharmacology,
B-2340 Beerse, Belgium

There is no doubt that dopamine-sensitive adenylate cyclase (D_1 site) is not involved in the antipsychotic effect of neuroleptic drugs. All the pharmacological and behavioural effects elicited by dopamine agonists and antagonists in the brain can only be explained if such an interaction occurs at the level of the dopamine receptor (D_2 receptor site); the D_1 site still remains in search of a function. Bovine parathyroid cells were reported to possess dopamine D_1 sites which should be involved in the control of parathormone secretion. However, the very poor pharmacological characterization and the lack of <u>in vivo</u> evidence do not allow to assess the dopaminergic nature of this hormone secretion. Dopamine-sensitive adenylate cyclase is thus not a receptor directly implicated in the dopaminergic neurotransmission; it is an enzyme which could have an important role in the control of long term metabolic effects such as the synthesis of neuronal constituents.

In the last decade, the term receptor has been used by so many people in so many different ways that we are, now, far from the original definition proposed by Langley ($\underline{1}$) in the early twentieth century. In fact, the receptor concept arose from physiological and pharmacological experiments; therefore, a physiological response is one of the most essential elements defining a receptor. According to Langley, a receptor is a site of competition for agonist and antagonist; the agonist produces a stimulus which leads to a physiological response and this is blocked by the antagonist. One can easily apply such a concept to the dopaminergic system; first it is necessary to clearly define the physiological effects of dopamine. Nausea, emesis,

0097–6156/83/0224–0022$06.00/0

stereotypy, hypermotility, decrease of prolactin secretion, stomach relaxation, neurogenic vasodilatation, antiparkinson effects, psychosis etc... are the main physiological or pharmacological responses to dopamine and its agonists. As a rule, these effects are measured under in vivo conditions: this is a prerequisite for calling those physiological responses. Sometimes one can detect a physiological effect in vitro as on isolated organs for instance; however to be relevant such an effect must correspond to a process occurring in the whole body.

One of the major sources of confusion around dopamine receptor originated with the idea that the increase of the cyclic AMP production by dopamine is a physiological effect of dopamine (2); starting from this viewpoint, it is not necessary to try to correlate the data obtained in vitro with physiological effects in vivo. In fact, the stimulation of cyclic AMP by dopamine is a biochemical effect for which one needs to find a physiological response in vivo. Just because an enzyme is stimulated or inhibited by a given neurotransmitter does not ipso-facto prove that such a process is involved in neurotransmission. As we will see further, numerous criteria must be fulfilled before an enzyme or a binding site may be called a receptor site.

The effects of dopamine quoted above, are antagonized by neuroleptic or antiemetic drugs (3,4) and all are mediated through the dopamine D_2 receptor site (5,6,7). This D_2 subtype (we prefer to call it the dopamine receptor) is the binding site labelled by dopamine antagonists like haloperidol and spiperone at nanomolar concentrations and by dopamine agonists at micromolar concentrations (7) and is not coupled to adenylate cyclase. More than 17 pharmacological, behavioural and biochemical parameters related to the effects of dopamine agonists and antagonists nicely correlate with IC_{50}-values obtained in the in vitro binding assay (5,8).

How the problem arises whether or not the dopamine-sensitive adenylate cyclase (D_1 site) (2) also answers these criteria or other criteria which justify it being called a dopamine receptor like the D_2 receptor site (5-8). The purpose of the present paper is to discuss this problem especially with regard to parathormone secretion. Special attention will be paid to the pharmacological characterization of this hormone secretion.

Is Dopamine-Sensitive Adenylate Cyclase a Dopamine Receptor ?

As recently quoted by Briggs and McAfee (9): "Rigorous quantitative pharmacology is required in equating the receptor utilized in synaptic transmission. Unfortunately the application of this pharmacology when available is often superficial". As a rule, a too small number of drugs, sometimes even one or two and often given at a single high concentration

have been used to characterize, pharmacologically, the
production of cyclic AMP stimulated by neurotransmitters. The
dopamine-sensitive adenylate cyclase did not escape this rule;
when Greengard's group reported the occurrence of
dopamine-sensitive adenylate cyclase in rat caudate nucleus as a
possible target for antipsychotic drugs (10,11), it rapidly
became evident that the potent neuroleptic drugs like
haloperidol and pimozide displayed a too low affinity for the
cyclase with regard to their high potency in pharmacological
tests and in the clinic. The discrepancy was most apparent for
pimozide which was found to be 10 times less active than
chlorpromazine on the cyclase whereas it is known to be 30 to 50
times more potent than chlorpromazine in vivo (3,4). Thereafter
certain drugs like sulpiride or domperidone were reported to be
practically inactive on the cyclase (5,12). Table I shows
clearly that numerous potent dopamine antagonists are poorly
active or inactive on the cyclase although they compete in the
binding assay, sometimes even at nanomolar concentration. Only
the phenothiazines and the thioxanthenes are approximatively
equiactive in both tests. It became obvious that the
antipsychotic effects of neuroleptic drugs were not mediated
through the cyclase (D_1 site) (5-8). There was a complete
lack of correlation between the inhibition of the
dopamine-sensitive adenylate cyclase and 17 behavioural
biochemical, pharmacological and clinical parameters (6,8). Two
other pieces of evidence indicate that the D_1 site is not
involved in dopaminergic neurotransmission in the brain. First
the in vivo accumulation of cyclic adenosine monophosphate
induced by apomorphine in the striatum was not blocked by
sulpiride and haloperidol whereas the behavioural effects were
blocked by both drugs (12). Secondly, when labelled
neuroleptics were injected into rats, the radioactivity was
found in association with the D_2 receptor, but never on the
D_1 sites (13); indeed the dopamine sensitive adenylate cyclase
and the binding site (D_2) possess a completely different
subcellular localization (14), a fact which gives rise to the
idea that they are two different entities not related to each
other.

From these considerations, one may conclude that the D_1
site is not directly implicated in dopaminergic
neurotransmission; this does not exclude a possible functional
role for the cyclase but hitherto it remains unknown. A
possible hypothesis is that the cyclase may control long term
metabolic effects such as the synthesis of neuronal constituents.

One may argue that the pharmacology of the D_1 site does
not necessarily have to be the same as that of the D_2 receptor
site; this is true but the problem is to get an in vivo
pharmacology for this D_1 site which entirely fits the data
obtained in vitro on the cyclase. Up to now there is no answer
to this problem. It is generally believed that parathormone

Table I
IC$_{50}$-values for various drugs on dopamine sensitive adenylate
cyclase (D$_1$) and ^{3}H-haloperidol binding (D$_2$)

| Drug | IC$_{50}$ (M) | | Ratio |
	D$_1$ site A	D$_2$ receptor site B	A/B
α-Flupenthixol	2.3 x 10^{-8}	2 x 10^{-8}	1.2
Chlorpromazine	1.1 x 10^{-6}	1.3 x 10^{-7}	8.5
(+)-Butaclamol	2 x 10^{-7}	1 x 10^{-8}	20
Haloperidol	7.5 x 10^{-7}	3.6 x 10^{-9}	208
Pimozide	1.5 x 10^{-5}	3.2 x 10^{-9}	4,687
Spiperone	2.2 x 10^{-6}	4.4 x 10^{-10}	5,500
Sulpiride	>10^{-3}	8 x 10^{-8}	> 10,000
Halopemide	>10^{-3}	1 x 10^{-8}	>100,000
Domperidone	2.4 x 10^{-4}	1.4 x 10^{-9}	142,857

secretion is mediated via the D$_1$ site (2); we will now examine
the criteria validating or invalidating this hypothesis.

Is Parathormone Secretion Mediated through a Dopamine D$_1$ Site ?

    In 1977, Brown et al. (15) reported that dopamine
(10^{-6} M) stimulates by 2-4 fold the secretion of parathormone
from dispersed bovine parathyroid cells.
    ADTN and other dopamine agonists mimicked this effect which
was antagonized by α- and ß-flupenthixol, the α-isomer being 100
times more potent. In a similar way, dopamine caused a rapid
20-30-fold increase in cellular cAMP in dispersed bovine
parathyroid cells. The potency of a series of dopaminergic
agonists and antagonists on adenylate cyclase activity
paralleled the effects of these ligands on cAMP accumulation and
parathormone secretion (16). It was concluded that bovine
parathyroid cells possess dopamine D$_1$ sites which are involved
in the control of parathormone secretion.
    This needs some comments; first, several other agents that
are not dopaminergic agonists, such as ß-adrenergic
catecholamines, secretin, phosphodiesterase inhibitors,
histamine, protaglandin (PGE$_2$) and cholera toxin were also
found to enhance cAMP accumulation and parathormone secretion in
human and bovine parathyroid cells (17,18,19) (cfr. Table II).
Moreover, calcium has been known for a long time, to play an
important role in the parathormone secretion; therefore the
effects of dopamine agonists are not selective or specific for a
given neurotransmitter.
    Somewhat surprising is the fact that dopamine is
ineffective in human dispersed parathyroid cells; this certainly

Table II

Compounds which can regulate (+) or not (-) cAMP production
and parathormone secretion in parathyroid cells

	Bovine parathyroid	Human parathyroid
Epinephrine	+	+
Dopamine	+	-
Histamine	-	+
Secretin	+	-
Prostaglandin	-	+
Phosphodiesterase inhibitors	+	-
Choleratoxin	+	?

limits the importance of this so-called D_1 dopamine site, more
especially as the present results do not allow the possibility
of a non specific effect of dopamine on parathormone secretion
to be excluded. One could assume, for instance that dopamine
like the other agonists can increase the membrane permeability
so that more hormone can be released after addition of these
compounds. Compatible with this hypothesis is the fact that the
increase of parathormone secretion elicited by dopamine is a
relatively slow process and that it can last as long as 60
minutes. As a rule the presence of agonists on a receptor site
for a long period of time leads to a desensitization phenomenon;
this is not the case here.

More intriguing is the fact that the parathormone secretion
occurs in the absence of dopamine; what dopamine is doing, is to
enhance a phenomenon already present; this is also at variance
with the normal physiological response to a neurotransmitter
which is generally an all or none process.

In fact the most important point concerns the very poor
pharmacological characterization of the cAMP formation enhanced
by dopamine and of the parathormone secretion. Firstly,
apomorphine is much less potent than dopamine, a fact which is
not compatible with what we know from pharmacological,
behavioural and even biochemical studies ([3],[4],[7]). It is
believed that apomorphine is a partial antagonist, but this has
never been found in in vivo conditions. The higher potency of
apomorphine is also reflected by its high affinity in
[3]H-haloperidol and ^{3}H-spiperone binding.

Secondly ergot derivatives which reveal a clearcut
agonistic activity on prolactin secretion and as antiparkinson
agents ([20]) were inactive on the cyclase. Surprisingly,
lisuride and lergotrile were found to be weak antagonists of
dopamine stimulated cAMP accumulation, but they could also
antagonize the cAMP production stimulated by isoproterenol; as

both compounds are completely inactive on the ß-receptors, this
finding constitutes a strong argument against the specificity of
the so-called dopaminergic and ß-adrenergic sites involved in
the cAMP accumulation in the bovine parathyroid cells. Another
point concerns the Ki's of l- and d-butaclamol on dopamine
sensitive adenylate cyclase which are 1 μM and 2.5 μM
respectively. It is well-known that the d-form is the active
enantiomer and l- the inactive one; therefore if d-butaclamol is
less active than l-butaclamol, and also that it is less active
than ß-flupenthixol, again an inactive enantiomer, the effects
of these drugs on the cyclase are thus irrelevant
physiologically. In fact to assess the dopaminergic nature for
the control of parathormone secretion, the following criteria
should be fulfilled:
1) a large number of dopamine antagonists should be tested
including compounds belonging to different chemical classes,
such as phenothiazines, butyrophenones, thioxanthenes,
diphenylbutylamines, domperidone, d- and l-butaclamol;
2) drugs belonging to other pharmacological classes should be
tested (α and ß-adrenergic, antiserotonergic, antihistamine,
anticholinergic, and lysosomotropic drugs like cloroquine for
instance;
3) the affinity of dopamine agonists should be compared to that
of non dopaminergic agonists;
4) a good correlation should be found between the inhibition of
the parathormone secretion measured under in vitro as well as in
vivo conditions and the decrease in cAMP and the inhibition of
dopamine-sensitive adenylate cyclase; in this regard, a large
number of drugs with a broad range of activity should be tested;
5) the increase of cAMP production elicited by dopamine
antagonists and its inhibition by dopamine antagonists should be
examined in the parathyroid under in vivo conditions.
     Only such an analytical approach can decide whether the
control of the parathormone secretion involves a dopaminergic
receptor.
     In my opinion, one may assume a priori that these criteria
will not be fulfilled; first, if the dopamine-sensitive
adenylate cyclase was really involved in the parathormone
secretion, the patients treated with neuroleptics and especially
with the most potent drugs on the D_1 sites (phenothiazine and
thioxanthene derivatives) would have normally revealed marked
changes in their parathormone secretion, just like as is the
case for the prolactin secretion; in fact such changes have
never been observed; secondly, a recent report clearly indicates
that the injection of dopamine in man does not modify
parathormone secretion although a marked decrease in prolactin
was observed (21). There is no receptor without physiological
response; the study of receptor requires a multidisciplinary
approach.
     Hitherto, it has not been proved that the D_1 site is

really involved in parathormone secretion; the D_1 site is thus an enzyme but not a receptor site since a physiological role has not been demonstrated.

Acknowledgments

Part of this work was supported by I.W.O.N.L. I thank David Ashton for his help in preparing the manuscript.

Literature Cited

1. Langley, J.N.   Proc. Roy. Soc. Ser. B 1906, 78, 107-194.
2. Kebabian, J.W.; Calne, D.B.   Nature 1979, 277, 93-96.
3. Niemegeers, C.J.E.; Janssen, P.A.J.   Life Sci. 1979, 24, 2201-2216.
4. Costall, B.; Naylor, R.J.   Life Sci. 1981, 28, 215-229.
5. Laduron, P.   Trends Pharmacol. Sci. 1980, 1, 471-474.
6. Laduron, P.   "Advances in Dopamine Research" Kohsaka, Ed.; Pergamon, 1982, Vol. 37; p. 71.
7. Seeman, P.   Pharmacol. Rev. 1980, 32, 229-313.
8. Laduron, P.   "Apomorphine and Other Dopaminomimetics" Gessa and Corsini, Eds.; Raven Press, 1981, Vol. 1; p. 85.
9. Briggs, C.A.; McAfee, D.A.   Trends Pharmacol. Sci. 1982, 3, 241-244.
10. Kebabian, J.W.; Petzold, G.L.; Greengard, P.   Proc. Natl. Acad. Sci. USA 1972, 69, 2145-2149.
11. Clement-Cormier, Y.C.; Kebabian, J.W.; Petzold, G.L.; Greengard, P.   Proc. Natl. Acad. Sci. USA , 71, 1113-1117.
12. Trabucchi, M.; Longoni, R.; Fresia, P.; Spano, P.F.   Life Sci. 1975, , 1551-1556.
13. Laduron, P.M.; Janssen, P.F.M.; Leysen, J.E.   Biochem. Pharmacol. 1978, 27, 323-328.
14. Leysen, J.E.; Laduron, P.   Life Sci. 1977, 20, 281-288.
15. Brown, E.M.; Carroll, R.; Aurbach, G.D.   Proc. Natl. Acad. Sci. USA 1977, 74, 4210-4213.
16. Brown, E.M.; Attie, M.F.; Reen, S.; Gardner, D.G.; Kebabian, J.; Aurbach, G.D.   Mol. Pharmacol. 1980, 18, 335-340.
17. Brown, E.M.; Hurwitz, S., Aurbach, G.D.   Endocrinology 1977, 100, 1696-1702.
18. Brown, E.M.; Gardner, D.G.; Windeck, R.A.; Aurbach, G.D.   Endocrinology 1979, 104, 218-224.
19. Brown, E.M.; Gardner, D.G.; Windeck, R.A.; Hurwitz, S.; Brennan, M.F.; Aurbach, G.D.   J. Clin. Endocrinol. Metab. 1979, 48, 618-626.
20. Schachter, M.; Bédard, P.; Debono, A.G.; Jenner, P.; Marsden, C.D.; Price, P.; Parkes, J.D.; Keenan, J.; Smith, B.; Rosenthaler, J.; Horowski, R.; Dorow, R.   Nature 1980, 286, 157-159.
21. Bansal, S.; Woolf, P.D.; Fischer, J.A.; Caro, J.F.   J. Clin. Endocrinol. Metab. 1982, 54, 651-652.

RECEIVED February 18, 1983

Dr. Brown's Replies to Dr. Laduron's Comments

Dr. Laduron raises a number of issues which require comment. He refers to the work of Langley (1) to provide support for a predominantly in vivo definition of receptors. By necessity, of course, these early experiments were based solely on physiological responses, such as muscle contraction. It is of interest, however, that Langley was prescient in postulating that interaction of agonists with a "receptive substance", which receives and transmits information, leads to a change in intracellular substances (?second messengers) and ultimately to a change in cellular function. Model systems such as bovine parathyroid cells and cells of the intermediate lobe of the rat pituitary gland have allowed for a direct demonstration of such a sequence. Dr. Laduron is correct that a physiological role for D-1 dopaminergic receptors has not yet been demonstrated in the brain, but, of course, this does not mean that such a function does not exist. A secretory response to dopamine has been demonstrated in vivo in the cow (2) which was not blocked by propranolol.

It should also be pointed out that much of the accumulated knowledge about well-defined receptor systems such as the beta-adrenergic receptor have come from models like the turkey or the frog erythrocyte, in which the physiological role of the receptor is unknown. By Dr. Laduron's definition, these systems do not have a beta receptor. This "shortcoming" has certainly not lessened the utility of these systems for carrying out detailed biochemical characterization of the beta receptor. Finally, a major weakness of a purely pharmacologic approach or even one which correlates binding sites with physiological responses is that it can be only correlative. With such an approach, it is always possible that if additional drugs were tested, exceptions would arise which would not support the postulated receptor-mediated linkage to a physiological response. A causal relationship between the interaction of an agonist with a receptor and a physiological response can only be established by working out the detailed molecular mechanisms by which the receptor-mediated changes in cellular function take place.

Dr. Laduron also feels that the dopamine-stimulated increase in cyclic AMP accumulation and PTH secretion may be "non-specific" and has not been shown to be mediated by a dopamine (D-1) receptor. We would simply like to reiterate the evidence we presented previously and to point out additional evidence for a specific receptor-mediated process. The effects of dopamine are rapid (less than a minute in vivo and nearly as fast in vitro) and are blocked specifically by dopaminergic antagonists of the phenothiazine, thioxanthine, and butyrophenone classes as well as by d-butaclamol (see below). They are totally unaffected, on the other hand, by the beta-adrenergic antagonist propranolol and the alpha-adrenergic blocker phentolamine at concentrations which totaly inhibit responses due to beta-adrenergic or alpha-adrenergic agonists, respectively. In addition, in pilot studies we have found that histamine, serotonin, octopamine and carbamyl choline have not consistent effect on cyclic AMP and/or PTH release. Finally, the effects of PGE_2 are not blocked by propranolol, alpha-flupenthixol, or phentolamine (3), the inhibitory effects of prostaglandin $F_{2\alpha}$ are not inhibited by phentolamine

(4), and the stimulatory effects of secretin are not inhibited by propranolol, alpha-flupenthixol, or phentolamine (5). Since methylisobutylxanthine and other phosphodiesterase inhibitors as well as cholera toxin act intracellularly through a non-receptor mediated process, we do not feel it likely that these agents act through any of the receptors noted above. We feel, therefore, that the specificity of the dopaminergic response of bovine parathyroid cells speaks for itself. We agree with Dr. Laduron, however, that the use of further drugs to test specificity would strengthen these data even further. It should be pointed out that the response of a cell to only a single class of agonists would be the exception rather than the rule. In the central nervous system, it is becoming increasingly clear that the interplay of a number of pharmacologic influences may determine the integrated response of given cell type.

To avoid confusion on the part of the reader, several inaccuracies on the part of Dr. Laduron should be pointed out.

(1) It is incorrect to state that desensitization does not occur in the dopaminergic response of bovine parathyroid cells. We did not, in fact, examine this point directly. Desensitization, however, probably actually does occur at more than one locus in this system. First, the secretory response of the parathyroid cell rapidly becomes refractory to agents such as dopamine and isoproterenol which produce large elevations in cyclic AMP (see ref. 2). Secondly, there is a progressive decrease in cellular cyclic AMP despite the continued presence of dopamine (see Figure 5 in our manuscript) possibly due to desensitization of the receptor-adenylate cyclase compex.

(2) Dr. Laduron uses the relative potency of d- and l-butaclamol as a major argument against the specificity of the parathyroid dopamine receptor. We apologize for the confusion which we apparently caused him in this regard. In our original manuscript, the symbol which Dr. Laduron took to indicate a potency of 1 micromolar for l-butaclamol actually referred to Footnote 1 which stated that l-butaclamol had no effect at 100 micromolar. In addition, because of the relatively low potency of d-butaclamol in our original studies (k_i = 2.5 micromolar), we carried out additional studies with fresh samples of d- and l-butaclamol. The potency of these newer samples are reflected in Table I of the present manuscript (k_i for d-butaclamol 5 $\simeq$ nanomolar; l-butaclamol again had no effect at 10^{-4} molar). Since we do not have data on the effects of the newer samples on adenylate cyclase, we have not included data for the effects of butaclamol on cyclase in the newer manuscript.

(3) Dr. Laduron states that both lisuride and lergotrile are antagonists at the beta-receptor in bovine parathyroid cells. He again uses this piece of evidence as a strong argument against the specificity of the dopamine receptor in this system. We know of no evidence, however, either in our work or that of others that lergotrile has such an effect on

parathyroid cells. Moreover, the k_i for lisuride at the beta-receptor differs by 3 to 4-fold from that for its effects on the dopamine receptor, suggesting that it is, in fact, acting at two different receptors.

(4) In Table II of his comments, Dr. Laduron points out differences between bovine and human parathyroid cells. Several of these are in error. Prostaglandins affect both human and bovine parathyroid cells. Phosphodiesterase inhibitors have not been tested directly in human parathyroid cells, although dibutyryl cyclic AMP, which may act in part by inhibiting phosphodiesterase, stimulated PTH release in fragments of human parathyroid glands (6).

(5) It is very likely an inaccuracy to call the D-1 dopamine receptor an enzyme. Because of the effects of guanine nucleotides on dopamine-stimulated adenylate cyclase, it is likely by analogy with other receptors (i.e. the β receptor) that the D-1 receptor is a distinct molecular entity which is coupled to adenylate cyclase by a guanine nucleotide binding subunit. Proof of this point, of course, will require physical separation of these entities.

(6) Dr. Laduron categorically states that the D-2 receptor is not coupled to adenylate cyclase. Recent work, however, utilizing both the effects of dopamine on the mammotroph as well as on dispersed cells of the intermediate lobe of the rat pituitary gland (the former of which Dr. Laduron appears to feel it is an example of the D-2 receptor) suggests that the D-2 receptor is, in fact, coupled to adenylate cyclase in a negative way. It may well turn out, therefore, that, like the alpha-2 adrenergic receptor, the D-2 receptor is coupled to adenylate cyclase through an inhibitory guanine nucleotide subunit. Whether or not other D-2 receptors in the central nervous system are linked to the cyclase in this fashion remains to be determined.

Literature Cited

1.   Langley, J.N. Proc. Roy. Soc. Ser. B. 1906, 78, 170-194.
2.   Blum, J.W.; Kunz, P.; Fischer, J.A.; Binswanger, U.; Lichtensteiger, W.; DaPrada, M. Am. J. Physiol. 1980, 239, E255.
3.   Gardner, D.B.; Brown, E.M.; Windeck, R.; Aurbach, G.D. Endocrinology 1978, 103, 577.
4.   Gardner, D.B.; Brown, E.M.; Windeck, R.; Aurbach, G.D. Endocrinology 1979, 104, 1.
5.   Windeck, R.; Brown, E.M.; Gardner, D.B.; Aurbach, G.D. Endocrinology 1978, 103, 2020.
6.   Dietel, M.; Dorn, G.; Montz, R.; Altenakr, E. Acta Endocrinologica (Kbh) 1977, 85, 541.

2

The D-2 Dopamine Receptor in the Intermediate Lobe of the Rat Pituitary Gland

Physiology, Pharmacology, and Biochemistry

J. W. KEBABIAN, M. BEAULIEU, T. E. COTE, R. L. ESKAY,
E. A. FREY, M. E. GOLDMAN, C. W. GREWE, M. MUNEMURA,
J. C. STOOF, and K. TSURUTA

National Institutes of Health, Experimental Therapeutics Branch, National
Institute of Neurological and Communicative Disorders and Stroke,
Bethesda, MD 20205

Dopaminergic neurons synapse upon the parenchymal
cells of the intermediate lobe (IL) of the rat
pituitary gland. Dopamine decreases the capacity
of the IL cells to synthesize cyclic AMP and
inhibits the release of αMSH and other peptides
from this tissue. The presence of a D-2 receptor
accounts for both of these phenomena. This D-2
dopamine receptor can be studied in a binding
assay using [3H]-spiroperidol, a dopamine
antagonist. These observations support the two
dopamine receptor hypothesis.

The hypothesis suggesting the existence of two categories
of dopamine receptor (1, Figure 1) arose from the observation
that lergotrile blocks the dopamine-induced enhancement of
striatal adenylate cyclase activity (2, see also Brown and
Dawson-Hughes, this volume) but mimicks the action of dopamine
upon the mammotrophs of the anterior pituitary gland (3). When
the hypothesis was initially put forward, the dopamine receptor
upon the mammotrophs was taken as the prototypic example of a
D-2 receptor. According to this hypothesis, stimulation of the
D-2 dopamine receptor does "not involve either the stimulation
of adenylate cyclase or the accumulation of intracellular
cyclic AMP" (1, 4-7). Although the mammotroph provided an
example of a D-2 dopamine receptor, the cellular heterogeneity
of the anterior pituitary gland limited the precision of this
biochemical model of the D-2 dopamine receptor (8; see also
Labrie et al., this volume). Indeed, many biochemical
investigations of the anterior pituitary dopamine receptor make
the untestable assumption that a biochemical signal not linked
to prolactin is generated by the mammotrophs and none of the
other cell types in the gland. The intermediate lobe (IL) of
the rat pituitary gland possesses a D-2 dopamine receptor
amenable to experimental investigation. The IL consists of an
homogeneous population of cells (9) possessing both a β-
adrenoceptor (10) and a dopamine receptor (11, 12). An

Name	D-1	D-2
Cyclase linkage	Yes	No
Location of proto-type receptor	Bovine parathyroid	Mammotroph of anterior pituitary
Dopamine	Agonist (μmolar potency)	Agonist (nmolar potency)
Apomorphine	Partial agonist or antagonist	Agonist (nmolar potency)
Dopaminergic ergots	Potent antagonist (nmolar potency) Weak agonist (μmolar potency)	Agonist (nmolar potency)
Selective antagonist	None known as yet	Metoclopramide sulpiride
Radiolabelled ligand	cis-flupenthixol	Dihydroergocryptine

Figure 1. Criteria for the classification of dopamine receptors.

Radiolabelled cis-flupenthixol can be used as a ligand specific for the dopamine receptor linked to adenylyl cyclase in the rat striatum (64). Its affinity for the dopamine receptor in the anterior pituitary has not been measured. (Previously (65) the two categories of dopamine receptors were designated as "α-dopaminergic" and "β-dopaminergic." This has led to confusion with the α and β adrenoreceptors. The new designations should prevent further confusion.)

understanding of the biochemical consequences of stimulation of the D-2 dopamine receptor has come from investigations of the IL dopamine receptor.

The Cell Biology of the Intermediate Lobe

The IL of the rat pituitary gland consists of a narrow band of cells adhering to the slightly larger neural lobe. The parenchymal cells of the IL synthesize proopiomelanocortin and then cleave this large molecule into smaller fragments which are converted into the hormones of the IL (14). In rodents, these hormones affect coat color by stimulating the dermal melanocytes to synthesize and secrete melanin (15, 16). The most widely studied of the melanotropic IL hormones is alpha-melanocyte-stimulating hormone (N-acetyl ACTH1-13 amide, αMSH). However, recent reports document that N-, O-diacetyl αMSH is the predominant molecular species in the rat IL (17,18, Goldman et al., submitted). In view of the hormonal control of coat color in rodents, the release of hormones from the rat IL provides a convenient, physiologically relevant sign of activity in the IL. The hormones released from the IL can be quantified by either bioassay or by radioimmunoassay. Several αMSH-like molecules are released from the IL; however, in the vast majority of studies, the amount of melanotrophic hormone or immunoreactive, αMSH-like material (IR-αMSH) is quantified in assays using αMSH as a standard.

Cyclic AMP and the Intermediate Lobe β-adrenoceptor

A β-adrenoceptor occurs upon the parenchymal calls of the rat IL and can be studied with biochemical procedures. The β - adrenoceptor itself can be identified with [125I]- monoiodohydroxybenzylpindolol (IHYP) (19). Using IHYP to identify binding sites and propranolol to define 'non-specific binding', specific binding sites for the radiolabeled ligand can be detected. The affinity of the specific binding site for a non-radioactive drug can be as determined by permitting the drug to compete with IHYP for occupancy of the specific binding sites. In the IL (as in many other tissues) the β-adrenoceptor regulates adenylate cyclase, the enzyme catalyzing the conversion of ATP into cAMP (19). The adenylate cyclase activity of cell-free homogenates of the IL is markedly stimulted (up to 7-fold) by catecholamines; the stimulatory effect of catecholamines requires the presence of GTP and can be blocked by several β-adrenergic antagonists (Figure 2, left; 19, 20)). Because the IHYP binding assay and the adenylate cyclase assay utilize similar assay conditions, a comparison between the affinities obtained in the two assays seems justified. The approximate agreement of the affinities from the two assays suggests that some or all of the binding sites are the receptors enhancing adenylate cyclase activity in the

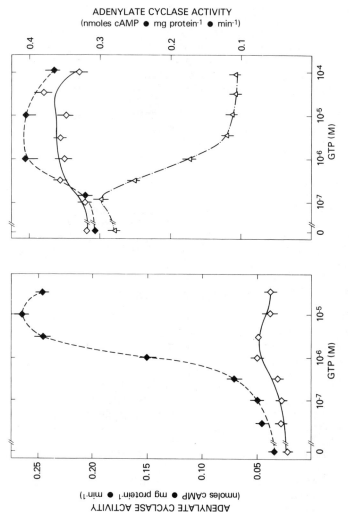

Figure 2. Guanosine 5'-triphosphate participation in the functioning of the IL β-adrenoceptor and the D-2 dopamine receptor in the IL of the rat pituitary gland.

Key: left, effects of GTP on basal and isoproterenol (♦)-stimulated adenylate cyclase activity in particulate material from a homogenate of fresh IL tissue; at the indicated concentrations, GTP was tested in the absence of drugs (◇) or in the presence of 3 μM isoproterenol (♦) (data are from Reference 20); and right, effect of GTP upon adenylate cyclase activity in a homogenate of cholera toxin-treated (30 nM, 2 h) IL tissue; at the indicated concentrations, GTP was tested alone (◇) or in combination with either 3 μM isoproterenol (♦) or 10 μM apomorphine (△) (data are from Reference 22).

cell-free homogenates. The β-adrenoceptor in the IL is a β$_2$-adrenoceptor (19, 21).
Cyclic AMP formed in response to stimulation of the β-adrenoceptor may initiate the intracellular events ultimately expressed as an enhanced release of αMSH. β-Adrenergic agonists (Figure 3) as well as other agents increasing cAMP content (e.g. theophylline, 3-isobuty-1-methylxanthine or cholera toxin) increase the release of αMSH (12, 22, 23). Furthermore, cyclic nucleotide analogues (e.g. dibutyryl cAMP or 8Br cAMP) also increase the release of αMSH (12, 24). A "working hypothesis" accounting for the effects of β-adrenergic agonists and other drugs upon the IL is as follows: occupancy of the β-adrenoceptor by an agonist enhances adenylate cyclase activity and thereby increases the intracellular content of cAMP. In turn, cAMP initiates the intracellular events leading to enhanced hormone secretion (Figure 4). The presence of calcium ions in the extracellular medium is obligatory for the stimulated release of αMSH (23, 25). At present the intracellular site(s) where calcium and cAMP impinge upon the release process is unknown.

The evidence from the IL, together with the evidence from many other biological systems, supports the hypothesis that cAMP mediates the effects of β-adrenergic agonists upon the IL. However, there are a number of unanswered questions about how this receptor, cAMP and calcium ions 'work' in the IL. For example, isoproterenol enhances the release of αMSH from dispersed IL cells at concentrations substantially lower than the concentrations required to either occupy the specific binding site identified with IHYP, stimulate adenylate cyclase activity in cell-free homogenates of the IL or enhance cAMP formation by intact IL cells (Figure 5, 19). At present no experimental observation upon IL tissue explains this apparent discrepancy. However, there are a number of indications that the IL gives a maximal physiological response while utilizing only a fraction of its capacity to synthesize cAMP. Thus, maximal isoproterenol-stimulated hormone release is accompanied by an accumulation of cAMP which is only 3% of the cyclic nucleotide accumulation elicited by a combination of cholera toxin and a phosphodiesterase inhibitor (23). Furthermore, the IL cells excrete cAMP into the extracellular medium so that after a few minutes of stimulation of the β-adrenoceptor, more cAMP is found outside the cells than is inside them. This evidence suggests that a minimal change in intracellular cAMP is needed to initiate secretion and points to the need for a greater understanding of calcium-dependent secretion, in general, before significant progress can be made in understanding the β-adrenergic stimulation of secretion from the IL, in particular.

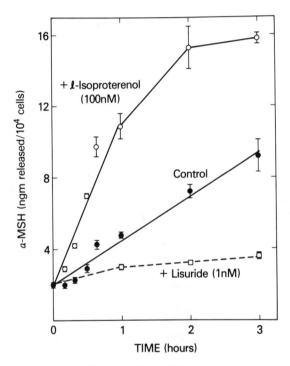

Figure 3. Effect of certain drugs on IR-αMSH release from rat cells. Dispersed rat IL cells were exposed to isoproterenol, lisuride, or no drug (control) for the indicated periods of time. Isoproterenol enhances the release of IR-αMSH while lisuride inhibits the release of IR-αMSH. (Data are from Reference 12.)

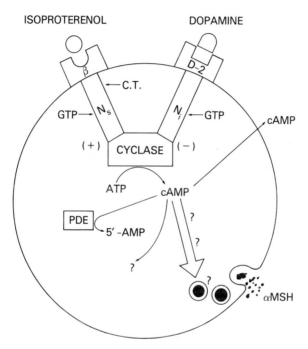

Figure 4. A representation of the "working hypothesis" of the dual regulation of adenylate cyclase activity and hormone release by a β-adrenoceptor and a D-2 dopamine receptor in the IL of the rat pituitary gland.

A β-adrenergic agonist (isoproterenol) occupies a β-adrenoceptor (β) to interact with a stimulatory guanyl nucleotide site (Ns); GTP interacts with Ns to stimulate adenylate cyclase (cyclase) activity. A dopaminergic agonist (dopamine) occupies a D-2 dopamine receptor (D-2) to interact with the hypothetical inhibitory guanyl nucleotide component (Ni); GTP interacts with Ni to inhibit adenylate cyclase activity. GTP activates Ns only in the presence of a β-adrenergic agonist and activates Ni only in the presence of a dopaminergic agonist. Cholera toxin selectively acts upon Ns. Increased formation of cAMP results in the enhanced release of IR-αMSH via an unknown mechanism(s). Newly formed cAMP can be hydrolyzed by intracellular phosphodiesterase(s) (PDE) to become 5'-AMP, or can be excreted into the extra-cellular milieu.

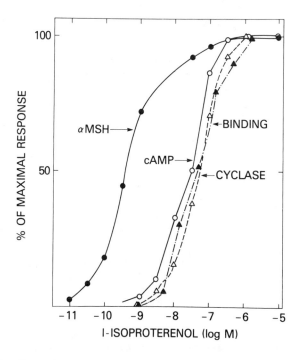

Figure 5. Comparison of the potency of isoproterenol in eliciting physiological and biochemical responses from the rat IL. Substantially lower concentrations of isoproterenol stimulate the release of IR-αMSH (αMSH) than are required to enhance cAMP accumulation by intact IL cells (cAMP), stimulate adenylate cyclase activity in cell-free homogenates of IL tissue (cyclase), or occupy the specific binding sites defined with IHYP (binding) (19).

Cyclic AMP and the Intermediate Lobe Dopamine Receptor

The rat IL is innervated by dopaminergic neurons. The
somata of these neurons are located in the arcuate nucleus;
their axons leave the brain via the pituitary stalk and project
to the intermediate lobe where they terminate in the vicinity
of the parenchymal IL cells (26). Baumgarten et al. describe
'synapse-like' connections between the dopaminergic nerve
terminals and the parenchymal cells of the IL (27). Several
lines of evidence support the view that dopaminergic
neurotransmission occurs within the IL. First, dopamine is the
predominant catecholamine within the IL (28, Saavedra, personal
communication). Second, radiolabelled dopamine is taken up and
stored in the IL; subsequently, this newly stored dopamine can
be released from the nerve terminals (29, 30). The available
evidence (which unfortunately does not include electrical
recording from the dopaminergic neurons themselves) suggests
that in vivo the dopaminergic neurons are spontaneously active
(30).

Dopaminergic Inhibition of Intermediate Lobe Adenylate
Cyclase. The IL dopamine receptor can be studied with
biochemical procedures. The receptor itself can be identified
in binding studies (31-33). Using [3H]-spiroperidol, a
dopamine antagonist from the butyrophenone series, to identify
binding sites and fluphenazine to define 'non-specific
binding', specific binding sites for the radiolabeled ligand
can be detected in the intermediate lobe (Figure 6). The rat
IL contains 20.2 fmole of high affinity (Kd=0.3 nM) specific
spiroperidol binding sites (33). The affinity of non-
radioactive dopaminergic agonists or antagonists can be
inferred from their ability to compete with [3H]-spiroperidol
for occupancy of the specific binding sites (e.g. see Figure
7). Lisuride is the most potent of the dopamine agonists
tested and spiroperidol is the most potent of the dopamine
antagonists tested.

The dopamine receptor in the IL regulates adenylate
cyclase activity. Stimulation of the IL dopamine receptor
decreases the capacity of the parenchymal cells to synthesize
cAMP (12, 19-22, 37). This effect of dopamine is especially
pronounced (and therefore amenable to experimental
investigation) when the IL adenylate cyclase activity has been
increased with either isoproterenol or cholera toxin. The
cholera toxin-treated IL has proven to be an especially useful
tissue for investigating dopaminergic drugs since their effect
upon the dopamine receptor can be segregated from any possible
effect upon the β-adrenoceptor (Figure 8). Demonstration of
the dopaminergic inhibition of IL adenylate cyclase activity
requires the presence of GTP (Figure 2, right).

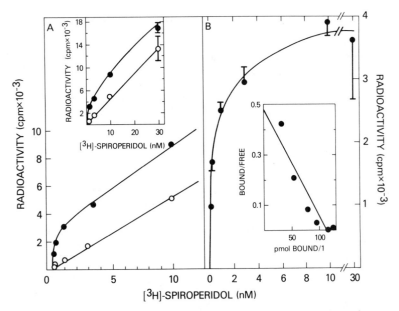

*Figure 6.   Binding of [3H]-spiroperidol to cell-free homogenates of the neuro-
intermediate lobe of the rat pituitary gland.*

Key: A, total (●) and nonspecific binding (i.e., binding in the presence of 2 μM fluphenazine)
(○) was determined in the presence of the indicated concentrations of [3H]-spiroperidol; and
B, specifically bound [3H]-spiroperidol (i.e., the difference between total and nonspecific binding) is shown as a function of the concentration of [3H]-spiroperidol. Left inset a plot of the
data in B according to Rosenthal's method: yields an apparent Kd of 0.2 nM and a maximal
concentration of specific binding sites of 94.1 fmol/mg protein (this is equivalent to 20.2
fmol/NIL) (33).

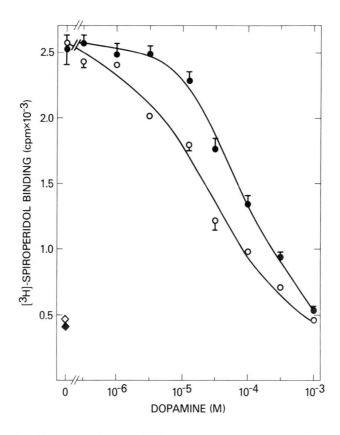

Figure 7. Competition between [3H]-spiroperidol and dopamine for occupancy of binding sites in the NIL of the rat pituitary gland. The amount of [3H]-spiroperidol bound to NIL tissue was determined in the presence of 2 μM fluphenazine (◇), 2 μM fluphenazine and 100 μM GTP (♦) or dopamine (at the indicated concentrations) in the absence (○) or presence (●) of 100 μM GTP (33).

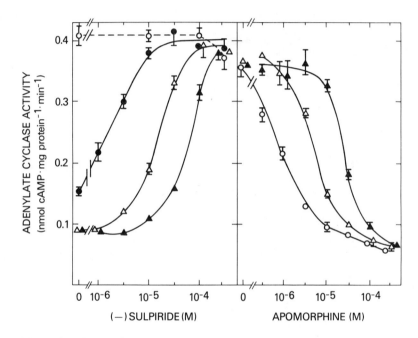

Figure 8. Competitive interaction between apomorphine and (—)-sulpiride in regulating adenylate cyclase activity in the IL (33).

IL tissue was treated with cholera toxin and adenylate cyclase activity was determined. Enzyme activity was determined in the presence of the indicated concentrations of (—)-sulpiride alone (○) or in combination with apomorphine (3 μM, ●; 10 μM, △; or 30 μM, ▲) (left). In a separate experiment, enzyme activity was determined in the presence of the indicated concentrations of apomorphine alone (○) or in combination with (—)-sulpiride (3 μM, △; 30 μM, ▲) (right). In both experiments, data represent the mean ± SE (n = 3).

[3H]-Spiroperidol Identified the Intermediate Lobe Dopamine Receptor. Because the spiroperidol binding assay and the adenylate cyclase assay are performed under equivalent assay conditions, a comparison between the affinities of drug in the two assay systems seems appropriate. Figure 9 shows the approximate agreement between the results obtained from the two experimental protocols. The similarity between the values obtained from the two assay systems supports the hypothesis that some or all of the specific [3H]-spiroperidol binding sites are the dopamine receptors which when stimulated inhibit adenylate cyclase activity. A "working hypothesis" of the biochemical organization of the D-2 dopamine receptor in the IL is as follows: the dopamine receptor on the exterior surface of an IL cell is coupled to adenylate cyclase by an inhibitory guanyl nucleotide regulatory protein (Ni), occupancy of the receptor by an agonist (i.e. stimulation) alters the properties of Ni so that intracellular GTP can interact with Ni and thereby inhibit the enzyme activity (Figure 4).

Dopamine affects three physiological processes in the IL which, with varying degrees of certainty, may be linked to the ability of dopamine to inhibit the capacity of the IL to synthesize cAMP. First the parenchymal cells of the rat IL display spontaneous propagated electrical spikes (38, 39). Dopamine diminishes the frequency, but not the amplitude, of these electrical discharges apparently by stimulating a D-2 dopamine receptor (40). The possible involvement of cyclic AMP in this inhibitory process has not been investigated. Second, dopamine inhibits the spontaneous release of IR-αMSH (Figure 3) or melanotrophic hormones from the IL (10). The presence of a receptor for dopamine in the IL of many vertebrate species has been inferred from the ability of dopamine to inhibit the release of melanotrophic hormones or IR-αMSH and the ability of dopaminergic antagonists to abolish this effect of dopamine (11, 12, 34). The dopamine-induced inhibition of adenylate cyclase activity and the dopamine-induced inhibition of electrical activity may participate in the dopaminergic inhibition of the release of IR-αMSH. Third dopamine will diminish the ability of β-adrenergic agonists to enhance either adenylate cyclase activity or the release of IR-αMSH (12, 35, 37). The dopamine-induced decrease in the β-adrenergic enhancement of cAMP synthesis seems likely to be involved in this phenomenon (38).

When the potency of dopaminergic agonists upon preparations of intact IL cells is compared to the potency of the same compounds in cell-free assay systems (i.e. adenylate cyclase or spiroperidol binding, a striking discrepancy emerges (Figure 10). For each of the agonists tested, the intact cells respond to approximately 1/100th the concentration of drug that is required to elicit a comparable effect from the cell-free homogenate. At the present time the hypothesis that the pituitary dopamine receptor can exist in two "states" of

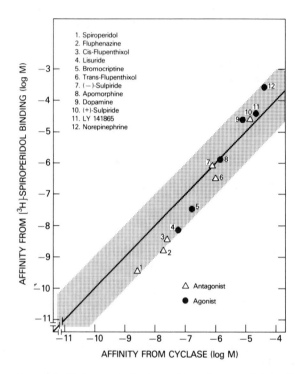

Figure 9.   Apparent affinity constants of agonists and antagonists for the dopamine receptor in the neurointermediate lobe of the rat pituitary gland.

In studies of [3H]-spiroperidol binding, the apparent affinity constant of an agonist or antagonist was determined on the basis of the ability of the compound to compete with [3H]-spiroperidol for occupancy of the specific binding site (IL contributes 86% of the specific [3H]-spiroperidol binding sites in the NIL) and assuming a competitive interaction between the radiolabelled ligand and the nonradioactive compound. In studies of adenylate cyclase activity the apparent affinity constant of an agonist was the concentration of agonist producing half of its maximal inhibition of the enzyme activity in cholera toxin treated IL tissue. The apparent affinity constant of each antagonist was determined on the basis of the ability of the compound to reverse the apomorphine-induced inhibition of adenylate cyclase activity, by assuming a competitive interaction between apomorphine and each antagonist. Each value is the mean of determinations from three or more separate experiments (33).

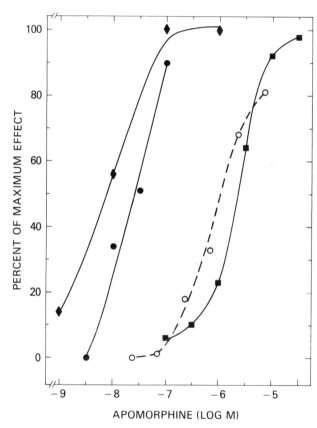

Figure 10.   Comparison of the potency of apomorphine as a dopaminergic agonist upon intact IL cells or cell-free homogenate of IL tissue.

For each response examined, inhibition of isoproterenol-stimulated cAMP accumulation by intact cells (♦); inhibition of basal release of IR-αMSH by intact cells (●); occupancy of specific [3H]-spiroperidol binding sites in a cell-free homogenate (○); and inhibition of isoproterenol-stimulated adenylate cyclase activity in a cell-free homogenate (■), the effect achieved with the indicated concentration of apomorphine is expressed as a percentage of the maximal effect of apomorphine (33).

differing affinity for agonists is the most attractive explanation for this discrepancy (41, 42; see also Caron, this volume).

Concluding Remarks

The IL of the rat pituitary gland possesses a dopamine receptor amenable to experimental investigation. Two interesting points have emerged from the combined physiological, pharmacological and biochemical approach towards this tissue. First, the physiological actions of dopamine upon the IL "seem not to involve either the stimulation of an adenylyl cyclase or the accumulation of intracellular cyclic AMP (1)." Furthermore, the IL dopamine receptor is stimulated by ergots (e.g. lisuride or bromocriptine) and blocked by the substituted benzamides (e.g. (-)-sulpiride). Therefore, the IL dopamine receptor ressembles the D-2 dopamine receptor in the classification schema of Kebabian and Calne (1). Second, the IL has become a useful component of a screening protocol designed to identify selective antagonists of either the D-1 or the D-2 receptor. The ability of LY 141865 (for structure see Figure 11) to selectively stimulate the D-2 receptor (36) and the ability of YM-09151-2 (Figure 11) to selectively block this receptor (43) were discovered in experiments utilizing the IL dopamine receptor as the model D-2 receptor and the goldfish retina as the model D-1 receptor. These compounds together with the selective D-2 agonists, RU 24213 and RU 24926, and the selective D-2 antagonist, domperidone, (for structures see Figure 11) provide useful pharmacological tools for identifying other D-2 receptors (44-48).

Two categories of dopamine receptor exist in the cardiovascular system. These two receptors have been designated as the DA-1 and the DA-2 receptors (49, see Goldberg and Kohli, this volume). The D-2 and the DA-2 receptors have very similar pharmacological properties. However, the DA-1 receptor and the D-1 receptor are perceived by Goldberg and his colleagues as being distinct entities. In particular, the DA-1 receptor is blocked by (+)-sulpiride; (+)-sulpiride is the most potent and most selective antagonist of the DA-1 receptor (49). In contrast, the D-1 receptor is at best only weakly antagonized by this compound (46, 47). Although (+)-sulpiride is the most potent and most selective antagonist of the DA-1 receptor in in vitro experiments, when it is tested in vitro upon strips of isolated canine mesenteric artery, it is a weak dopamine antagonist (50). Similarly, (+)-sulpiride is a weak antagonist of the D-1 receptor in the bovine parathyroid gland (see Brown and Dawson-Hughes, this volulme). In the future, it will be important to identify the basis for the difference between the in vivo and in vitro potencies of sulpiride. However, in the context of the present discussion about the number and properties of the cardiovascular dopamine receptors,

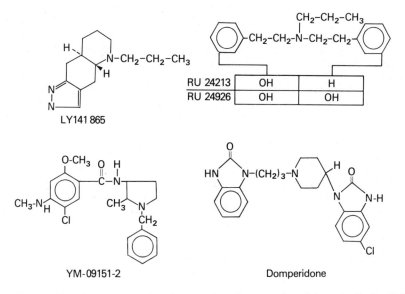

Figure 11. Structures of agonists (top) and antagonists (bottom) of the D-2 receptor.

it seems worthwhile to consider four similarities between the
DA-1 and the D-1 receptors. First, (+)-bulbocapnine is an
antagonist capable of blocking either the D-1 or the DA-1
receptor; this compound is substantially less potent as an
antagonist of the DA-2 receptor and is devoid of antagonist
activity upon the D-2 receptor (51, 52, Itoh, Goldman and
Kebabian, unpublished observations). Second, SKF 38393 is an
agonist upon the D-1 and the DA-1 receptor, yet it does not
elicit the physiological responses occurring as a consequence
of stimulation of either the D-2 or the DA-2 receptor (53).
Third, LY 141865, a selective D-2 agonist (36), fails to
stimulate the DA-1 receptor in the renal vasculature of the rat
(54). Fourth, domperidone is an antagonist of either the D-2
or the DA-2 receptor which does not block either the D-1 or the
DA-1 receptors (45, 48). These four similarities between the
D-1 and the DA-1 receptor suggest that they are extremely
similar if not identical in terms of their pharmacological
properties. Clearly, it will also be helpful to determine the
basis for the difference in the in vivo and in vitro potencies
of the substituted benzamides. However, at present it may be
premature to discount the striking similarities between the D-1
and the DA-1 receptors.

The IL also serves as a model for dopaminergic
neurotransmission in the brain; the concepts first elucidated
in the IL can be applied to the brain. For example, in the
neostriatum the interaction between dopamine and a D-2 receptor
inhibits the stimulated efflux of cAMP (and by inference the
enhancement of adenylate cyclase activity) occurring as a
consequence of stimulation of the D-1 receptor 55, 56). The
types of dopamine receptors in the CNS and their physiological
function are discussed in this volume by Stoof.

Literature Cited

1. Kebabian, J. W.; Calne, D. B. Nature 1979, 277, 93-96.
2. Kebabian, J. W.; Calne, D. B.; Kebabian, P. R. Commun. in
    Psychopharmacol. 1977, 1, 311-318.
3. Kebabian, J. W. "Dopamine, Advances in Biochemical
    Psychopharmacology Volume 19"; Raven Press: New York,
    1978; pp 131-154.
4. Borgeat, P.; Chavancy, G.; Dupont, A.; Labrie, F.;
    Arimura, A.; Schally, A. V. Proc. Natl. Acad. Sci. USA
    1972, 69, 2677-2681.
5. DeCamilli, P.; Macconi, D.; Spada, A. Nature 1979, 278,
    252-254.
6. Giannattasio, G.; De Ferrari, M. E.; Spada, A. Life Sci.
    1981, 28, 1605-1612.
7. Thorner, M. O.; Hackett, J. T.; Murad, F.; MacLeod, R. M.
    Neuroendocrinology 1980, 31, 309-402.

8.   Baker, B. "Handbook of Physiology: Section 7, Volume 4, Part I, The Pituitary Gland and Its Neuroendocrine Control"; American Physiological Society: Bethesda, Maryland, 1974; pp 45-80.
9.   Dube, D.; Lissitzky, J. C.; Leclerc, R.; Pelletier, G. Endocrinology 1978, 102, 1283-1291.
10. Bower, A.; Hadley, M. E.; Hruby, V. J. Science 1974, 184, 70-72.
11. Morgan, C. M.; Hadley, M. E. Neuroendocrinology 1976, 21, 10-19.
12. Munemura, M.; Eskay, R. L.; Kebabian, J. W. Endocrinology 1980, 106, 1795-1803.
13. Baker, B. I. J. Endocrinol. 1974, 63, 533-538.
14. Eiper, B. A.; Mains, R. E. Endocrine Rev. 1980, 1, 1-27.
15. Silvers, W. K. "The Coat Colors of Mice"; Springer-Verlag: New York, 1979; pp 1-5.
16. Geschwind, I. I.; Huseby, R. A.; Nishioka, R. Rec. Prog. Horm. Res. 1972, 28, 91-130.
17. Rudman, D.; Chawla, R. K.; Hollins, B. M. J. Biol. Chem. 1979, 254, 10102-10108.
18. Browne, C. A.; Bennett, H. P. J.; Solomon, S. Biochemistry 1981, 20, 4538-4546.
19. Cote, T.; Munemura, M.; Eskay, R. L.; Kebabian, J. W. Endocrinology 1980, 107, 108-116.
20. Cote, T. E.; Grewe, C. W.; Kebabian, J. W. Endocrinology 1982, 110, 805-811.
21. Muniere, H. and Labrie, F. Eur. J. Pharmacol. 1982, 81, 411-420.
22. Cote, T. E.; Grewe, C. W.; Tsuruta, K.; Stoof, J. C.; Eskay, R. L.; Kebabian, J. W. Endocrinology 1982, 110, 812-819.
23. Tsuruta, K.; Grewe, C. W.; Cote, T. E.; Eskay, R. L.; Kebabian, J. W. Endocrinology 1982, 110, 1133-1139.
24. Baker, B.I. J. Endocrinol. 1974, 63, 533-538.
25. Tomiko, S. A.; Taraskevich, P. S.; Douglas, W. W. Neuroscience 1981, 6, 2259-2267.
26. Lindvall, O.; Bjorklund, A. "Handbook of Experimental Pharmacology, Volume 9"; Plenum Press: New York, 1978; pp 139-231.
27. Baumgarten, H. G.; Bjorklund, A.; Holstein, A. F.; Nobin, A. Z. Zellforsch. 1972, 126, 483-517.
28. Saavedra, J. M.; Palkovits, M.; Kizer, J. S.; Brownstein, M.; Zivin, J. J. Neurochem. 1975, 25, 257-260.
29. Tilders, F. J. H.; Mulder, A. H.; Smelik, P. G. Neuroendocrinology 1975, 18, 125-130.
30. Tilders, F. J. H.; Van Der Woude, H. A.; Swaab, D. F.; Mulder, A. H. Brain Res. 1979, 171, 425-435.
31. Sibley, D. R.; Creese, I. Endocrinology 1980, 107, 1405-1419.

32. Stefanini, E.; Devoto, P.; Marchisio, A. M.; Vernaleone,
     A. M.; Collu, R. Life Sci. 1980, 26, 583-587.
33. Frey, E. A.; Cote, T. E., Grewe, C.W.; Kebabian, J. W.
     Endocrinology 1982, 110, 1897-1904.
34. Munemura, M.; Cote, T. E.; Tsuruta, K., Eskay, R. L.;
     Kebabian, J. W. Endocrinology 1980, 197, 1676-1683.
35. Cote, T. E.; Grewe, C. W.; Kebabian, J. W. Endocrinology
     1981, 108, 420-426.
36. Tsuruta, K.; Frey, E. A.; Grewe, C. W.; Cote, T. E.;
     Eskay, R. L.; Kebabian, J. W. Nature 1981, 292, 463-465.
37. Muniere, H.; Labrie, F. Life Sci. 1982, 30, 963-968.
38. Douglas, W. W.; Taraskevich, P. S. J. Physiol. (London)
     1978, 285, 171-184.
39. Douglas, W. W.; Taraskevich, P. S. J. Physiol. (London)
     1980, 209, 623-630.
40. Douglas, W. W.; Taraskevich, P. S. J. Physiol. (London)
     1982, 326, 201-211.
41. De Lean, A.; Kilpatrick, B. F.; Caron, M. G. Endocrinology
     1982, 110, 1064-1066.
42. Sibley, D. R.; De Lean, A.; Creese, I. J. Biol. Chem.
     1982, 257, 6351-6361.
43. Grewe, C. W.; Frey, E. A.; Cote, T. E.; Kebabian, J. W.
     Eur. J. Pharmacol. 1982, 81, 149-152.
44. Euvard, C; Ferland, L.; Dipaolo, T.; Beaulieu, M.; Labrie,
     F.; Oberlander, C.; Raynaud, J.P.; Boissier, J.R.
     Neuropharmacology, 1980, 19, 379.
45. Denef, C.; Follebouckt, J.-J. Life Sci. 1978, 23, 431.
46. Watling, K.J.; Dowling, J.E.; Iversen, L.L. Nature, 1979,
     281, 578.
47. Watling, K.J.; Dowling, J.E. J. Neurochem. 1981, 35, 559.
48. Kohli, J.D.; Glock, D.; Goldberg, L.I.   Eur. J.
     Pharmacol. 1983, in press.
49. Goldberg, L.I.; Kohli, J.D.   Commun. Psychopharmacol.
     1979, 3, 447-456.
50. Kohli, J.D.; Takeda, H.; Ozaki, N.; Goldberg, L.I.
     "Advances in the Biosciences: Volume 20, Peripheral
     Dopaminergic Receptors"; Pergamon Press: Oxford, U.K.,
     1979; pp 143-149.
51. Goldberg, L.I.; Musgrave, G.E.; Kohli, J.D. "Sulpiride
     and Other Benzamides"; Italian Brain Research Foundation
     Press: Milan, Italy, 1979; pp 73-81.
52. Shepperson, N.B.; Duval, N.; Massingham, R.; Langer, S.Z.
     J. Pharmacol. Exp. Ther. 1980, 221, 753-761.
53. Setler, P.E.; Sarau, H.M.; Zirkle, C.L.; Saunders, H.L.
     Eur. J. Pharmacol. 1978, 50, 419.
54. Hahn, R.A.; McDonald, B.; Martin, M. J. Pharmacol. Exp.
     Ther. 1983, 224, 206-214.
55. Stoof, J. C.; Kebabian, J. W. Nature 1981, 294, 366-368.
56. Stoof, J. C.; Kebabian, J. W. Brain Res. 1982, 250, 263-
     270.

RECEIVED November 4, 1982

Commentary: The DA_ Dopamine Receptor in the Anterior and Intermediate Lobes of the Pituitary Gland

F. LABRIE, H. MEUNIER, M. GODBOUT, R. VEILLEUX, V. GIGUERE, L. PROULX-FERLAND, T. DI PAOLO, and V. RAYMOND

Le Centre Hospitalier de l'Université Laval, Department of Molecular Endocrinology, Quebec G1V 4G2, Canada

Dopamine released from nerve endings of tuberoinfun-dibular neurons is the main factor of hypothalamic origin which exerts a direct inhibitory effect on prolactin secretion in the anterior pituitary gland. The dopamine receptor on pituitary mammotrophs is negatively coupled to adenylate cyclase. Estrogens inhibit while androgens and progestins as well as androgens facilitate the activity of dopamine on prolactin secretion by an action at a post-receptor site. The peptidic corticotropin-releasing factor (CRF) is a potent stimulator of adenylate cyclase activity and peptide secretion in pars intermedia cells. The intermediate lobe is thus controlled by two stimulatory receptors, namely CRF and β-adrener-gic while it is inhibited by a dopaminergic receptor negatively coupled to adenylate cyclase. The nomen-clature DA_+, DA_- and DA_0 is proposed for dopamine receptors which are positively, negatively or un-coupled, respectively, to adenylate cyclase. The dopamine receptors in both the anterior and interme-diate lobes of the pituitary gland are negatively coupled to adenylate cyclase and are thus typical of the DA_ type.

The predominant role of dopamine in normal brain functions as well as in diseases such as Parkinson's disease and schizophrenia is a strong stimulus for research on the dopamine receptor and its mechanism of action. Since dopamine appears to be the main if not the exclusive factor of hypothalamic origin acting as inhibitor of prolactin secretion (1, 2) and changes of prolactin secretion can be measured with precision in vivo and in vitro, the anterior pituitary gland is an attractive model for a dopaminergic recep-tor. Another model of great interest is the intermediate lobe of the pituitary gland, a pure population of dopamine-sensitive cells

0097–6156/83/0224–0053$06.00/0

specialized in the secretion of proopiomelanocortin-related peptides, specially α-MSH (α-melanotropin) and β-endorphin (3). Both
mammotrophs of the rat anterior pituitary gland and pars intermedia cells can easily be grown in monolayer culture where changes
of dopamine receptors can be correlated with biochemical parameters such as changes in cyclic AMP levels and peptide secretion
into the medium.

Kebabian et al. (this volume) present a clear and precise review of the large body of biochemical data obtained by their group
on the control of pars intermedia cell activity by β-adrenergic
agents and dopamine during the last years (4–13). Since the peptidic corticotropin-releasing factor (CRF) is a potent stimulator of
pars intermedia cell activity (14, 15, 16), we will complement
this review by summarizing recent data on the mechanism of action
of CRF in the intermediate lobe as well as its interaction with β-
adrenergic and dopaminergic agents. Data will also be presented
on the coupling of the anterior pituitary dopamine receptor controlling prolactin secretion with adenylate cyclase as well as on
the marked interactions of this receptor with gonadal steroids.
Finally, taking into account all available evidence, we will pro-
pose to use the nomenclature DA_+, DA_- and DA_0 for dopamine receptors coupled positively, negatively or uncoupled, respectively,
to adenylate cyclase. The dopamine receptors in the anterior and
intermediate lobes of the pituitary gland are examples of dopamine
receptors negatively coupled to adenylate cyclase (DA_- type).

The dopamine receptor in the anterior pituitary gland is negative-ly coupled to adenylate cyclase

The role of cyclic AMP as modulator of prolactin secretion
was first suggested by the finding of a stimulatory effect of
cyclic AMP derivatives (17–22) and inhibitors of cyclic nucleotide
phosphodiesterase activity such as theophylline and IBMX (22–26)
on the secretion of this hormone. More convincing evidence
supporting a role of cyclic AMP in the action of dopamine on prolactin secretion had to be obtained, however, by measurement of
adenohypophysial adenylate cyclase activity or cyclic AMP accumulation under the influence of the catecholamine. As illustrated
in Fig. 1, addition of 100 nM dopamine to male rat hemipituitaries
led to a rapid inhibition of cyclic AMP accumulation, a maximal
effect (30% inhibition) being already obtained 5 min after addition of the catecholamine. Thus, while dopamine is well known to
stimulate adenylate cyclase activity in the striatum (27, 28), its
effect at the adenohypophysial level in intact cells is inhibitory. Dopamine has also been found to exert parallel inhibitory
effects on cyclic AMP levels and prolactin release in ovine adenohypophysial cells in culture (29) and purified rat mammotrophs
(30). Using paired hemipituitaries obtained from female rats, Ray
and Wallis (22) have found a rapid inhibitory effect of dopamine
on cyclic AMP accumulation to approximately 75% of control.

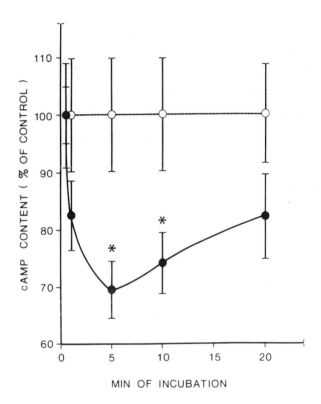

Figure 1. Time course of the effect of dopamine (100 nM) on cyclic AMP accumulation in male rat anterior pituitaries (54). Key: O—O, control; and ●—●, with dopamine.

As much as higher concentrations of dopamine are required to stimulate striatal adenylate cyclase activity as compared to the affinity of the drug for the agonistic sites of the receptor, the sensitivity of the anterior pituitary adenylate cyclase to dopamine agonists in both human prolactinoma (26) and rat adenohypophysial homogenate (31) is lower than the potency of the compounds to inhibit prolactin secretion in intact cells (2). This variable loss of sensitivity upon cell homogeneization is a phenomenon frequently observed in other systems (32).

The reports that DA did not show significant changes of cyclic AMP levels in the adenohypophysis (33, 34, 35) have led to the suggestion that the pituitary DA receptor is a prototype of D_2 receptors (D_2) not coupled to adenylate cyclase (36). These negative results were however obtained using a very heterogeneous tissue, the total anterior pituitary gland. The background of cyclic AMP contributed by the five other cell types could easily mask the changes in cyclic AMP occurring specifically in mammotrophs under the influence of dopamine. Using pituitary homogenate obtained from female rats containing a higher proportion of mammotrophs than the tissue used in previous studies (33, 34), Giannattasio et al. (31) could demonstrate a clear GTP-dependent inhibition of adenylate cyclase activity.

Since cyclic AMP derivatives and inhibitors of cyclic nucleotide phosphodiesterase stimulate prolactin release (17, 37, 38) and dopamine is a potent inhibitor of prolactin secretion (1, 2, 39), it is not surprising that the catecholamine does not stimulate the adenylate cyclase system. On the contrary, the data summarized above show that the pituitary DA receptor is negatively coupled to adenylate cyclase. The pituitary DA receptor is thus a typical DA_-receptor (40, 41). In view of the multiplicity of factors involved in the control of prolactin secretion, including sex steroids, it is likely that mechanisms other than cyclic AMP are involved (39, 42). It does however appear that inhibition of cyclic AMP formation by dopamine is a key element in a multifactorial control system responsible for the fine tuning of prolactin secretion.

Modulation of the anterior pituitary dopamine receptor by gonadal steroids

Administration of estrogens is well known to cause an increase in plasma prolactin (PRL) levels in man (43, 44) as well as in the rat (45-48). This stimulatory effect of estrogens is also observed in vitro in anterior pituitary gland explants (49), tumoral adenohypophysial cells (50) and normal rat anterior pituitary cells in primary culture (39, 40, 42, 51). Seventeen-β-estradiol (E_2) does not only stimulate basal and TRH-induced PRL secretion in rat anterior pituitary cells in culture but it can also reverse almost completely the inhibitory effect of dopamine (DA) agonists on PRL release (40).

Our data have shown that the non-aromatisable androgen DHT
and progestins can act directly at the pituitary level at physiological concentrations to inhibit spontaneous PRL secretion and
reverse the well-known stimulatory effect of E_2 ($\underline{39}$, $\underline{40}$, $\underline{42}$, $\underline{51}$).
In addition, the effect of E_2, DHT and progestins is observed, not
only on spontaneous and TRH-induced PRL secretion, but also in the
presence of IBMX. This last observation suggests that the marked
modulatory effects of sex steroids are exerted at a step following
cyclic AMP formation.

These results prompted us to suggest a model for the site of
the modulatory role of sex steroids on PRL secretion in the female
rat (Fig. 2). TRH probably acts through activation of adenylate
cyclase, the effect of TRH being mimicked by cyclic AMP derivati-
ves and inhibitors of cyclic nucleotide phosphodiesterase ($\underline{17}$, $\underline{52}$,
$\underline{53}$) (Fig. 2) while DA receptors are negatively coupled to adenylate cyclase ($\underline{26}$, $\underline{54}$). Estrogens exert their stimulatory action on
TRH receptors as well as at steps following cAMP formation while
androgens and progestins apparently inhibit PRL secretion at a
post-receptor step. This schematic representation does not exclude
other elements which are probably involved in the control of the
mammotroph cell but it only indicates the main sites of steroid
action using the presently available evidence. Ca^{2+} and the phos-
phatidylinositol response are also intimately involved in the
secretion mechanisms in mammotrophs.

The present data indicate that besides providing a better
knowledge of the control of prolactin secretion, the tuberoinfun-
dibular system can also be used as a model for other less accessible dopaminergic systems in the central nervous system. The interest of such studies in the rat on the interaction of sex steroids
with brain catecholaminergic systems is strengthened by the observation that estrogen administration to male or female patients
receiving neuroleptics can facilitate the appearance of parkinsonian symptoms ($\underline{55}$). Moreover, treatment with estrogens has been
found to improve the symptoms of tardive dyskinesias induced by
chronic treatment with L-Dopa or neuroleptics ($\underline{56}$). The above-
mentioned effects of estrogen treatment in the human are compatible with an antidopaminergic action and open the possibility of
clinical applications of sex hormones in the treatment of neurological as well as psychiatric diseases.

The dopamine receptor in the intermediate lobe of the pituitary
gland is negatively coupled to adenylate cyclase

As illustrated in Fig. 3A, dopamine leads to a 30% (p < 0.01)
inhibition of basal cyclic AMP levels in pars intermedia cells at
an ED_{50} value of 5.0 nM. An almost identical potency of dopamine
is observed on the elevated cyclic AMP concentration induced by
simultaneous incubation with 30 nM (-)isoproterenol (Fig. 3B).
Similar inhibitory effects of dopamine are observed in the presence of a phosphodiesterase inhibitor, isobutylmethylxanthine, thus

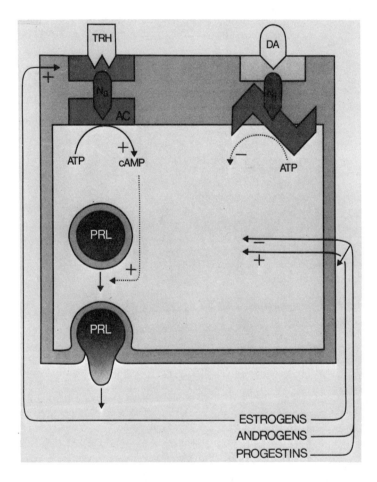

Figure 2.   Representation of the site of interaction of sex steroids, TRH, and dopamine on PRL secretion in the anterior pituitary gland according to the data obtained in the female rat.

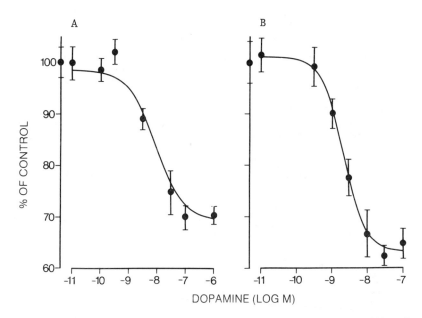

Figure 3. *Effect of increasing concentrations of dopamine on basal (A) and (−)isoproterenol-induced (B) cyclic AMP levels in rat pars intermedia cells in culture. Cells were incubated for 30 min in DMEM containing 5 mM HEPES, 100 μM ascorbic acid, and the indicated concentrations of dopamine alone (A). In B, 30 nM (−)isoproterenol was present during the last 4 min of incubation. Data are expressed as percent of control (in the absence of dopamine). Control cyclic AMP levels were 0.96 ± 0.10 and 5.29 ± 0.13 pmol/2 × 10⁵ cells in the absence (A) or presence (B) of 30 nM (−)isoproterenol, respectively (41).*

indicating that the changes in cyclic AMP levels are due to parallel inhibition of adenylate cyclase activity rather than to stimulation of cyclic nucleotide phosphodiesterase activity. Apomorphine, the prototype of dopaminergic agonists (57), is approximately 3 times more potent than dopamine in inhibiting pars intermedia cyclic AMP levels at a K_D value of 1.5 nM (41). Ergot
alkaloids which act as potent dopaminergic agonists on prolactin
release in the anterior pituitary gland (2) also show potent agonistic activity in the intermediate lobe: lisuride, 2-bromo-α-
ergocryptine, pergolide, dihydroergocryptine and lergotrile inhibit basal cyclic AMP levels at K_D values of 0.05, 0.20, 0.27,
0.50 and 2.7 nM, respectively. (-)Norepinephrine and (-)epinephri-
ne are less potent, half-maximal inhibitory effects on cyclic AMP
levels being measured at 30 and 50 nM, respectively. Propranolol
(100 nM) was present during incubation with (-)epinephrine and
(-)norepinephrine in order to block their interaction with the β-
adrenergic receptor which is stimulatory on cyclic AMP accumulation in pars intermedia cells (4).
    The specificity of the dopamine receptor was further studied
with a series of dopaminergic antagonists of well known pharmacological activity. The 30-40% inhibitory effect of 10 nM dopamine
was completely reversed by the addition of increasing concentrations of the potent neuroleptics (+)butaclamol (K_D = 1.5 nM) and
(-)sulpiride (K_D = 0.5 nM) while their pharmacologically weak
enantiomers (-)butaclamol and (+)sulpiride were 86 and 167 times
less potent, respectively. The neuroleptics spiroperidol, thio-
properazine, domperidone, haloperidol, fluphenazine and pimozide
completely reversed the inhibitory effect of dopamine at low K_D
values ranging from 0.02 to 0.8 nM (41).

Coupling of dopamine receptors with adenylate cyclase: DA$_+$, DA$_-$
and DA$_0$ receptors

    Dopamine can thus be added to the list of hormones and neuro-
transmitters which can stimulate or inhibit cyclic AMP formation,
depending upon their tissue of action. Thus, while dopamine stimulates cyclic AMP formation in parathyroid cells, superior cervical
ganglia, retina and striatal tissue (27, 58-61), it inhibits the
accumulation of the cyclic nucleotide in cells of the intermediate
and anterior lobes of the pituitary gland. Opposite effects on the
cyclic AMP system are also found with LHRH which stimulates and
inhibits cyclic AMP levels in the anterior pituitary gland (62)
and ovary (63), respectively. Similarly, alpha-adrenergic agents
show opposite effects on cyclic AMP formation in brain (64) and
platelets (65). PGE$_1$ stimulates cyclic AMP formation in the anterior pituitary gland (62) while it inhibits the same parameter in
fat cells (66).
    There is much ambiguity in the currently used classifications
of dopamine receptors. This complexity is partly due to the combined use of a classification based on binding data (67, 68) and

another one based on function of the dopamine receptor (36, 41, 54). Until more precise data are available in a large variety of tissues (using homogeneous cell populations, if possible), classifications of dopamine receptors based on the binding characteristics of various ligands and classifications describing functional coupling to various cellular activities such as activation of adenylate cyclase, should be kept separate. Although clinical interest has stimulated the synthesis of long series of dopamine agonists and antagonists, there is a relatively small number of model systems where the binding characteristics of dopaminergic agents can be correlated with a typical profile of cellular function. In fact, precise parameters of intact cellular activity under dopaminergic control can be measured with precision only in a few dopaminergic systems, namely prolactin-secreting (2), pars intermedia (4) and parathyroid (60) cells. Unfortunately, turning and stereotypic behavior in rats, psychiatric and neurological symptoms in man and other dopaminergic functions of the central nervous system are relatively imprecise.

While the classification of α-adrenergic receptors into α_1- and α_2 (69, 70) and β-adrenergic receptors into β_1 and β_2-subtypes (71, 72) is based on the rank order of potency of a series of catecholamines and analogs to elicit a large series of biological responses, there is no such large scale basis for classification of dopamine receptors. One of the earliest biochemical actions of dopamine to have been described is the stimulation of striatal adenylate cyclase activity. Although it has the disadvantage of a highly heterogenous cell population and its assay is performed in a broken cell preparation, striatal dopamine-induced adenylyl cyclase has been much useful for characterization of the dopamine receptor (27, 58, 73). However, not all dopaminergic responses appear to be mediated by cyclic AMP (74, 75, 76). This has led to the proposal by Kebabian and Calne (36) of two categories of dopamine receptors, namely D_1 (linked to adenylate cyclase) and D_2 (not linked to cyclase) (Table 1, 36). The parathyroid and the pituitary mammotroph were taken as examples of D1 and D2 receptors, respectively. While few data were then available on the coupling of the anterior pituitary dopamine receptor to adenylate cyclase (26, 29), it is now quite clear that the D1-D2 terminology can no longer adequately describe the functional coupling between the dopamine receptors and adenylate cyclase in various tissues.

It has in fact been convincingly demonstrated that the anterior pituitary dopamine receptor is negatively coupled to adenylate cyclase in both tumoral (26), and normal (54) adenohypophysial tissue. Thus, the adenohypophysis cannot be taken, as originally proposed (36), as example of a receptor not coupled to adenylate cyclase. Moreover, it is well known that some dopaminergic responses appear independent of cyclic AMP (74, 75, 76). In order to take into account all the available data, it appears preferable to use the terminology DA_+, DA_- and DA_0 (54, Fig. 4). Other advantages of this terminology are its clear separation from the nomen-

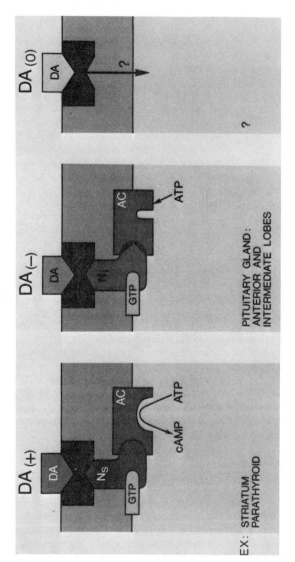

Figure 4.   Representation of the classification of the dopamine receptor based on its coupling with adenylate cyclase activity. DA_+ receptors (left) are coupled to adenylate cyclase through the Ns GTP-binding protein (91) with secondary activation of adenylate cyclase. DA_- receptors (middle) are coupled through the Ni GTP-binding protein, thus resulting in inhibition of cyclic AMP formation. DA_0 receptors (right) are those uncoupled to cyclic AMP formation, the example being possibly some autoreceptors on nigrostriatal dopaminergic neurons.

clature D_1 to D_4 based on binding data (68), the use of the well-accepted acronym DA for dopamine and the lack of possible confusion with the vitamin D receptor.

Corticotropin-releasing factor stimulates adenylate cyclase in the intermediate lobe of the pituitary gland

Although the first evidence suggesting the presence of hypothalamic substances controlling pituitary gland activity was that of a corticotropin-releasing factor (CRF) (77, 78), it is only recently that the structure of ovine CRF could be elucidated (79, 80). The 41-amino acid peptide is a potent stimulator of adrenocorticotropin (ACTH) secretion in vivo in the rat as well as in adenohypophysial cells in culture (79, 81, 82, 83). Somewhat unexpectedly, we have recently found that the administration of CRF leads to a parallel increase in plasma levels, not only of ACTH, a peptide secreted mainly by the anterior pituitary gland, but also of α-melanocyte-stimulating hormone (α-MSH) (14), a secretory product originating almost exclusively from the intermediate lobe of the pituitary gland (3, 84). In order to obtain direct evidence that CRF is acting directly on cells of the pars intermedia rather than indirectly at the hypothalamic level, we have studied the effect of CRF on α-MSH secretion in pars intermedia cells in culture.

After a 6-h incubation with pars intermedia cells in culture, a maximal concentration of ovine CRF (300 nM) causes a 2-fold stimulation of α-MSH release at an ED_{50} value of 1 nM (Fig. 5B). It can also be seen that preincubation with the potent glucocorticoid dexamethasone under conditions which lead to an almost complete inhibition of ACTH secretion in corticotrophs of the anterior pituitary gland (85), has no inhibitory effect on either spontaneous or CRF-induced α-MSH secretion.

In rat pars intermedia cells, the rate of secretion of the proopiomelanocortin-derived peptides (α-MSH being the major secretory product) (3, 85) was so far known to result from a balance between the stimulatory effect of β-adrenergic agents and the inhibitory influence of dopaminergic substances (14, 41, 86-89). The present data clearly demonstrate that in addition to β-adrenergic agents, a second substance, namely CRF, could well be involved as physiological stimulator of the activity of pars intermedia cells.

Since the adenylate cyclase system has been well demonstrated to play a mediatory role in controlling the action of β-adrenergic and dopaminergic agents in pars intermedia cells (5-13, 41), we have studied the possibility of a similar role of cyclic AMP in . CRF action. After a 10-min incubation with increasing CRF concentrations, an approximately 6-fold increase in cyclic AMP content is measured at an ED_{50} value of 6 nM (Fig. 5A). A maximal stimulatory effect of CRF on cyclic AMP accumulation is observed 2 min after addition of CRF. As observed on α-MSH secretion, preincuba-

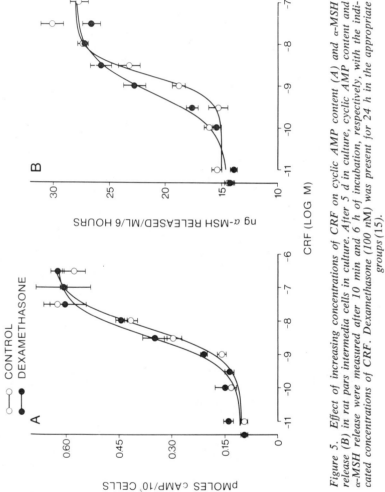

Figure 5. Effect of increasing concentrations of CRF on cyclic AMP content (A) and α-MSH release (B) in rat pars intermedia cells in culture. After 5 d in culture, cyclic AMP content and α-MSH release were measured after 10 min and 6 h of incubation, respectively, with the indicated concentrations of CRF. Dexamethasone (100 nM) was present for 24 h in the appropriate groups (15).

tion with dexamethasone has no inhibitory effect on basal or CRF-induced cyclic AMP accumulation.

The above-described data show that CRF added to cells of the rat intermediate lobe in culture causes a rapid stimulation of α-MSH release and cyclic AMP accumulation, thus demonstrating a direct action of the peptide on pars intermedia cells (15). It is however difficult, using intact cells, to dissociate between increases in cyclic AMP levels due to stimulation of adenylate cyclase activity or to inhibition of cyclic nucleotide phosphodiesterase or to a combination of both effects. Definitive proof of the role of adenylate cyclase in the action of CRF in the intermediate lobe of the pituitary gland is provided by the following findings of a CRF-induced stimulation of adenylate cyclase activity in homogenate of rat and bovine pars intermedia cells.

As illustrated in Fig. 6, maximal concentrations of synthetic ovine CRF cause a 100% stimulation of adenylate cyclase activity in bovine pars intermedia particulate fraction at an ED_{50} value of 150 nM. Since guanyl nucleotides have been demonstrated to play a central role in mediating the activation of adenylate cyclase by many hormones (90, 91), we next investigated the possibility of such a role of GTP in CRF action. We thus studied the interaction of CRF with GTP and the two previously known regulators of pars intermedia cell activity using the β-adrenergic agonist (-)isoproterenol and dopamine.

As illustrated in Fig. 7, 3 μM CRF and 1 μM (-)isoproterenol cause a 190 and 110% stimulation of adenylate cyclase activity in rat pars intermedia particulate fraction, respectively. An additive effect is observed when both stimulatory agents are present. Dopamine (30 μM), on the other hand, has no significant effect alone. However, in the presence of GTP, the catecholamine causes a 40 to 60% inhibition of adenylate cyclase activity stimulated by CRF, ISO or CRF + ISO. It can also be seen that while 0.3 mM GTP alone causes a 100% increase in basal adenylate cyclase activity, it leads to a marked potentiation of the effect of ISO and CRF on [^{32}P] cyclic AMP accumulation. It should be noticed that in the absence of the guanyl nucleotide, dopamine has no inhibitory effect on adenylate cyclase activity in any of the groups studied.

The present data clearly demonstrate that the 41-amino acid ovine CRF is a potent stimulator of adenylate cyclase activity in rat and bovine pars intermedia tissue. Our previous data have shown that CRF causes a rapid and marked stimulation of cyclic AMP accumulation in rat pars intermedia cells in culture (15). The final proof of the role of adenylate cyclase in the observed changes of cyclic AMP levels had to be obtained by direct measurement of adenylate cyclase activity.

As mentioned earlier, guanyl nucleotides have been found to play an important role in the activation of adenylate cyclase activity by many hormones (90, 91). The present observations show that in pars intermedia tissue, GTP causes an almost doubling of the stimulatory effect of CRF while that of the β-adrenergic ago-

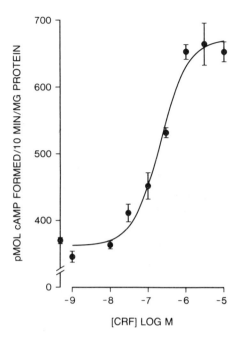

Figure 6. *Effect of increasing concentrations of CRF on adenylate cyclase activity in bovine pars intermedia pituitary homogenate.*

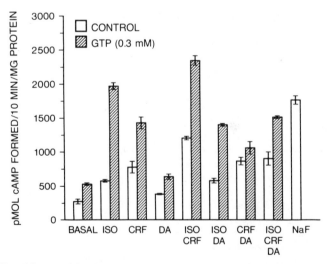

Figure 7. Effect of 0.3 mM GTP on basal, 1 μM (+)isoproterenol (ISO)-, 3 μM CRF-, 30 μM dopamine (DA), CRF + ISO-, ISO + DA-, CRF + DA-, ISO + CRF + DA, and 10 mM NaF-induced adenylate cyclase activity in rat pars intermedia homogenate. Cells of the intermediate lobe were separated from the posterior lobe by mechanical agitation, homogenized in 25 vol (w/v) of 10 mM Tris-HCl (pH 7.6)–2 mM EGTA. The 40,000 × g pellet was resuspended in the same buffer and used at the concentration of 35–40 μg/assay tube.

nist isoproterenol is increased by 3-fold.   It is also of great
interest that the guanyl nucleotide is required for the inhibitory
effect of dopamine on isoproterenol, as well as on CRF-induced
adenylate cyclase activation.   Such data suggest that CRF activation of adenylate cyclase activity in the rat pars intermedia
involves the participation of a GTP-regulatory protein (90, 91).
Moreover, the present data support the proposal that the dopamine
receptor in the intermediate lobe of the pituitary gland is negatively coupled to adenylate cyclase and is thus of the DA_type
(41, 54).   While the present observations leave little doubt about
the role of the adenylate cyclase system as mediator of the action
of CRF, the possible involvement of other intracellular mechanisms
remains to be determined.

The above-described findings indicate that CRF should now be
considered, in addition to β-adrenergic agents, as stimulator of
the activity of pars intermedia cells (Fig. 8).   Although the
physiological importance of the stimulatory action of CRF on the
activity of the intermediate lobe of the pituitary gland remains
to be determined, it suggests an unrecognized role of α-MSH and
possibly of other proopiomelanocortin-derived peptides secreted by
pars intermedia cells in the defense mechanisms accompanying the
response to stress.

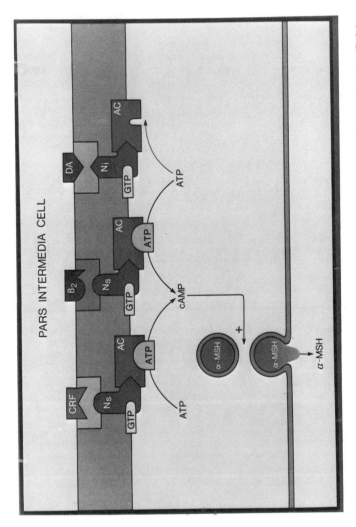

Figure 8. Representation of the interaction between CRF, β-adrenergic, and dopaminergic (DA.) receptors in the control of pars intermedia cell activity. The CRF and β-adrenergic receptors stimulate adenylate cyclase activity through interaction with the Ns–GTP-binding component. Dopamine, on the other hand, interacts with the Ni–GTP-binding component, causing inhibition of basal as well as CRF- and β-adrenergic-induced adenylate cyclase activity.

Literature Cited

1. MacLeod, R.M. "Frontiers in Neuroendocrinology" Martini, L.,
   Ganong, W.F., Eds.: Raven Press, New York, 1976; p. 169.
2. Caron, M.G.; Beaulieu, M.; Raymond, V.; Gagné, B.; Drouin, J.;
   Lefkowitz, R.J. J. Biol. Chem. 1978, 253, 2244-53.
3. Lissitzky, J.C.; Morin, O.; Dupont, A.; Labrie, F.; Seidah,
   N.G.; Chretien, M.; Lis, M.; Coy, D.H. Life Sci. 1978, 22,
   1715-22.
4. Munemura, M.; Eskay, R.L.; Kebabian, J.W. Endocrinology 1980,
   106, 1795-803.
5. Munemura, M.; Côté, T.E.; Tsuruta, K.; Eskay, R.L.; Kebabian,
   J.W. Endocrinology 1980, 107, 1676-83.
6. Cote, T.; Munemura, M.; Eskay, R.L.; Kebabian, J.W.
   Endocrinology 1980, 107, 108-16.
7. Cote, T.E.; Grewe, C.W.; Kebabian, J.W. Endocrinology 1981,
   108, 420-6.
8. Cote, T.E.; Grewe, C.W.; Kebabian, J.W. Endocrinology 1982,
   110, 805-11.
9. Cote, T.E.; Grewe, C.W.; Tsuruta, K.; Stoof, J.C., Eskay,
   R.L.; Kebabian, J.W. Endocrinology 1982, 110, 812-9.
10. Tsuruta, K.; Frey, E.A.; Grewe, C.W.; Cote, T.E.; Eskay, R.L.;
    Kebabian, J.W. Nature 1981, 282, 463-5.
11. Tsuruta, K.; Grewe, C.W.; Cote, T.E.; Eskay, R.L.; Kebabian,
    J.W. Endocrinology 1982, 110, 1133-9.
12. Frey, E.A.; Cote, T.E.; Grewe, C.W.; Kebabian, J.W.
    Endocrinology 1982, 110, 1897-904.
13. Grewe, C.W.; Frey, E.A.; Cote, T.E.; Kebabian, J.W. Eur. J.
    Pharmacol. 1982, 81, 149-52.
14. Proulx-Ferland, L.; Labrie, F.; Dumont, D.; Cote, J.; Coy,
    D.H.; Sueiras, J. Science 1982, 217, 62-3.
15. Meunier, H.; Lefèvre, G.; Dumont, D.; Labrie, F. Life Sci.
    1982, 31, 2129-35.
16. Labrie, F.; Gagné, B.; Lefèvre, G.; Meunier, H. Science 1982,
    (in press).
17. Lemay, A.; Labrie, F.; FEBS Lett. 1972, 20, 7-10.
18. Nagasawa, H.; Yanai, R. J. Endocrinol. 1972, 55, 215-6.
19. Samli, M.; Lai, M.F. Endocrinology 1973, 83, 767-76.
20. Dannies, P.S.; Gautvick, K.M.; Tashjian, A.J. Jr.
    Endocrinology 1976, 93, 1147-59.
21. Hill, M.K.; MacLeod, R.M.; Orcutt, P. Endocrinology 1976, 99,
    1612-7.
22. Ray, K.P.; Wallis, M. Biochem. J. 1981, 194, 119-28.
23. Parsons, J.A.; Nicoll, C.C. Fed. Proc. 1970, 29, abst. 750.
24. Kimura, H.; Calabro, M.A.; MacLeod, R.M. Fed. Proc. 1976, 35,
    abst. 305.
25. Nakano, H.; Fawcett, C.P.; McCann, S.M. Endocrinology 1976,
    98, 278-88.
26. De Camilli, P.; Macconi, D.; Spada, A. Nature 1979, 278, 252-
    4.

27. Kebabian, J.W.; Petzold, G.L.; Greengard, P. Proc. Natl. Acad. Sci. USA 1972, 69, 2145-9.
28. Kebabian, J.W. "Advances in Cyclic Nucleotide Research" (Greengard P, Robison G.A. Eds.), Raven Press, New York, vol. 8, 1973; p. 421.
29. Adams, T.F.; Wagner, T.O.; Sawyer, H.R.; Nett, T.M. Biol. Reprod. 1979, 21, 735-47.
30. Barnes, G.D.; Brown, B.L.; Gard, T.G.; Atkinson, D.; Ekins, R.P. J. Mol. Cell. Endocrinol. 1978, 12, 273-84.
31. Giannattasio, G.; de Ferrari, M.E.; Spada, A. Life Sci. 1981, 28, 1605-12.
32. Brown, E.M.: Gardner, D.G; Windeck, R.A.; Aurbach, G.D. Endocrinology 1978, 103, 2323-33.
33. Schmidt, M.J.; Hill, L.E. Life Sci. 1977, 20, 789-98.
34. Stefanini, E.; Devoto, P.; Marchisio, A.M.; Vernaleone, A.M.; Collu, R. Life Sci. 1980, 26, 583-87.
35. Spano, P.F.; Govoni, S.; Trabucchi, M. Adv. Biochem. Psychopharmacol. 1977, 19, 155-66.
36. Kebabian, J.W.; Calne, D.B. Nature 1979, 277, 93-6.
37. Labrie, F.; Pelletier, G.; Lemay, A.; Borgeat, P.; Barden, N.; Dupont, A.; Savary, M.; Coté, J.; Boucher, R. "Karolinska Symposium on Research in Reproductive Endocrinology", Diczfalusy, E. (ed.), Geneva, 1973; p. 301.
38. Pelletier, G.; Lemay, A.; Béraud, G.; Labrie, F. Endocrinology 1972, 91, 1355-71.
39. Labrie, F.; Beaulieu, M.; Caron, M.G.; Raymond, V. "Progress in Prolactin Physiology and Pathology" (Robyn, C.; Harter, M. Eds), Elsevier-North Holland Biomedical Press, 1978; p. 121.
40. Labrie, F.; Ferland, L.; Denizeau, F.; Beaulieu, M. J. Steroid Biochem. 1980, 12, 323-30.
41. Meunier, H.; Labrie, F. Life Sci. 1982, 30, 963-8.
42. Raymond, V.; Beaulieu, M.; Labrie, F.; Boissier, J.R. Science 1978, 200, 1173-5.
43. Frantz, A.G.; Kleinberg, D.L.; Noel, G.L. Rec. Progr. Horm. Res. 1972, 28, 527-90.
44. Yen, S.S.C.; Ehara, Y.; Siler, T.M. J. Clin. Invest. 1974, 53, 652-5.
45. Ajika, K.; Krulich, L.; Fawcett, C.P.; McCann, S.M. Neuroendocrinology 1972, 9, 304-15.
46. Chen, L.; Meites, J. J. Endocrinol. 1970, 86, 503-5.
47. De Léan, A.; Garon, M.; Kelly, P.A.; Labrie, F. Endocrinology, 1977, 100, 1505-10.
48. De Léan, A.; Ferland, L.; Drouin, J.; Kelly, P.A.; Labrie, F. Endocrinology 1977, 100, 1496-504.
49. Nicoll, C.S.; Meites, J. Endocrinology 1962, 70, 272-6.
50. Haug, E.; Gautvik, K.M. Endocrinology 1976, 99, 1482-9.
51. West, B.; Dannies, P.S. Endocrinology, 1980, 106, 1108-13.

52. Labrie, F.; Borgeat, P.; Lemay, A.; Lemaire, S.; Barden, N.;
    drouin, J.; Lemaire, I.; Jolicoeur, P.; Bélanger, A. "Advances
    in Cyclic Nucleotide Research" (Drummond, P.; Greengard, P.;
    Robinson, G.A. Eds.), Raven Press, New York, vol. 1, 1975; p.
    787.
53. Rose, J.C.; Conklin, P.M. Proc. Soc. Exp. Biol. Med. 1978,
    158, 524-9.
54. Labrie, F.; Borgeat, P.; Barden, N.; Godbout, M.; Beaulieu,
    M.; Ferland, L.; Lavoie, M. "Polypeptide Hormones", (Beers,
    R.F.; Bassett, E.G., Eds.), Miles Int. Symp. Series no. 12,
    Raven Press, New York, 1980; p. 235.
55. Gratton, L. Union Méd. Canada 1960, 89, 681-94.
56. Bédard, P.; Langelier, P.; Villeneuve, A. Lancet 1977, 2,
    1367-8.
57. Anden, N.E.; Rubenson, A.; Fuxe, K.; Hökfelt, T. J. Pharmac.
    Pharmacol. 1967, 19, 627-9.
58. Kebabian, J.W.; Greengard, P. Science 1971, 174, 1346-9.
59. Brown, J.H.; Makman, M.H. Proc. Natl. Acad. Sci. USA 1972, 69,
    539-43.
60. Brown, E.M.; Carroll, R.J.; Aurbach, G.D. Proc. Natl. Acad.
    Sci. USA 1977, 74, 4210-13.
61. Brown, F.M.; Matties, M.F.; Reen, S.; Gardner, D.G.; Kebabian,
    J.; Aurbach, G.D. Mol. Pharmacol. 1980, 18, 335-40.
62. Borgeat, P.; Chavancy, G.; Dupont, A.; Labrie, F.; Arimura,
    A.; Schally, A.V. Proc. Natl. Acad. Sci. USA 1972, 69, 2677-
    81.
63. Massicotte, J.; Veilleux, R.; Lavoie, M.; Labrie, F. Biochem.
    Biophys. Res. Commun. 1980, 84, 1362-6.
64. Chasin, M.; Rivkin, I.; Mamrak, F.; J. Biol. Chem. 1971, 246,
    3037-41.
65. Salzman, E.W.; Neri, L.L. Nature 1969, 224, 609-10.
66. Fain, J.N.; Jacobs, M.D.; Clement-Cormier, Y.C. Pharmacol.
    Rev. 1973, 25, 67-118.
67. Sokoloff, P.; Martres, M.P.; Schwartz, J.C. Naun-Sch. Arch-
    Pharmacol. 1980, 315, 89-102.
68. Seeman, P. Pharmacol. Rev. 1980, 32, 229-313.
69. Langer, S.L. Biochem. Pharmacol. 1974, 23, 1793-800.
70. Berthelsen, S.; Pettinger, W.A. Life Sci. 1977, 21, 595-606.
71. Lands, A.M.; Arnold, A.; McAuliff, J.P.; Luduena, F.P.; Brown,
    T.F. Nature 1967, 214, 597-98.
72. Lands, A.M.; Luduena, F.P.; Buzzo, H.J. Life Sci. 1976, 6,
    2241-49.
73. Clement-Cormier, Y.C.; Kebabian, J.W.; Petzold, G.L.;
    Greengard, P. Proc. Natl. Acad. Sci. USA 1974, 71, 1113-7.
74. Aghajanian, G.K.; Bunney, B.S. Adv. Biochem. Psychopharmacol.
    1977, 16, 433-8.
75. Kruger, B.K.; Forn, J.; Walters, J.R. Molec. Pharmacol.
    1976, 12, 639-48.
76. DiChiara, G.; Porceddu, M.L.; Spano, P.F.; Brain Res. 1977,
    130, 374-82.

77. Saffran, M.; Schally, A.V. J. Biol. Chem. 1955, 33, 408-15.
78. Guillemin, R.; Rosenberg, B. Endocrinology 1955, 57, 599-607.
79. Vale, W.; Spiess, J.; Rivier, C.; Rivier, J. Science 1981, 213, 1394-7.
80. Spiess, J.; Rivier, J.; Rivier, C.; Vale, W.; Proc. Natl. Acad. Sci. USA 1981, 78, 6517-21.
81. Giguère, V.; Labrie, F.; Côté, J.; Coy, D.H.; Sueiras-Diaz, J.; Schally, A. Proc. Natl. Acad. Sci. USA 1982, 79, 3466-9.
82. Rivier, C.; Brownstein, M.; Spiess, J.; Rivier, J.; Vale, W. Endocrinology 1982, 110, 272-8.
83. Labrie, F.; Veilleux, R.; Lefevre, G.; Coy, D.H.; Sueiras-Diaz, J.; Schally, A.V. Science 1982, 216, 1007-8.
84. Dubé, D.; Lissitzky, J.C.; Leclerc, R.; Pelletier, G. Endocrinology 1978, 102, 1283-91.
85. Raymond, V.; Lepine, J.; Lissitsky, J.; Coté, J.; Labrie, F. Mol. Cell. Endocrinol. 1979, 16, 113-22.
86. Meunier, H.; Labrie, F. Eur. J. Pharmacol. 1982, 81, 411-420.
87. Howe, A. J. Endocrinol. 1973, 59, 385-409.
88. Hadley, M.E.; Hruby, V.J.; Bower, A. Gen. Comp. Endocrinol. 1975, 26, 24-35.
89. Tilders, F.J.H.; Mulder, A.H.; Smelik, P.G. Neuroendocrinology 1975, 18, 125-30.
90. Rodbell, M.; Birnbaumer, L.; Pohl, S.L.; Krans, H.M.L. J. Biol. Chem. 1971, 246, 1877-82.
91. Rodbell, M. Nature 1980, 284, 17-21.

RECEIVED November 4, 1982

3

The Dopamine Receptor of the Anterior Pituitary Gland

Ligand Binding and Solubilization Studies

M. G. CARON, B. F. KILPATRICK, and A. DE LEAN[1]

Duke University Medical Center and Howard Hughes Medical Institute Research Laboratories, Departments of Physiology and Medicine, Durham, NC 27710

In the anterior pituitary gland dopamine inhibits the release of prolactin. The receptor for dopamine in the porcine anterior pituitary gland has been studied by direct ligand binding using [^{3}H]n-propylapomorphine, an agonist or [^{3}H]spiroperidol, an antagonist. The membrane-bound receptor appears to exist in two different forms which are reciprocally favored by agonists and antagonists. K_D values of agonists for these two forms of the receptor differ by 30-200 fold for various agonists but only 2- to 10-fold for [^{3}H]spiroperidol. Guanine nucleotides (GTP; EC_{50} = 20-30 μM) convert the form of the receptor having high affinity for agonists and low affinity for antagonists almost completely to a form having low affinity for agonists and high affinity for antagonists. This effect of guanine nucleotides can be mimicked qualitatively and quantitatively by treatment of membranes with either N-ethylmaleimide (EC_{50} = 6 μM) or heat (53° for 4 min.) without affecting the total number of receptors as measured by [^{3}H]spiroperidol binding. Treatment of membrane preparations with the detergent digitonin (1-2%) leads to the solubilization of up to 30% of the total receptor sites with retention of their ability to bind ligands with a dopaminergic specificity. Solubilization, however, results in a loss of the nucleotide sensitivity of the receptor. Labelling of membranes with [^{3}H]n-propylapomorphine prior to solubilization

[1] Current address: Clinical Research Institute of Montreal, 110 West Pine Avenue, Montreal, Quebec H2W 1R7, Canada

yields a soluble agonist-receptor complex whose
affinity can be modified by guanine nucleotides.
These results suggest the involvement of another
component, presumably a guanine nucleotide
binding protein, in the mechanism of interaction
of dopaminergic ligands with the dopamine
receptor of the porcine anterior pituitary gland.

The diverse physiological effects of catecholamines in
mammalian tissues are mediated by specific receptors. The
effect of epinephrine and norepinephrine are mediated by beta-
and alpha-adrenergic receptors while those of dopamine are
elicited by distinct receptor sites. In the last decade,
dopamine receptors have been identified as being involved in
the modulation of the enzyme adenylate cyclase. The original
observations of Kebabian et al. ($\underline{1}$) that a dopamine stimulated
adenylate cyclase was present in the central nervous system was
one of the first indications as to a possible biochemical
mediator for dopaminergic responses. More recently, several
reports have suggested that a pharmacologically distinct
subtype of dopamine receptor might be coupled to an attenuation
of the effector, adenylate cyclase, in some tissues ($\underline{2,3,4}$).
The classification proposed by Kebabian and Calne ($\underline{5}$) of two
different subtypes termed D_1 and D_2 was based on
pharmacological differences between sites identified by ligand
binding and sites mediating dopamine stimulation of adenylate
cyclase. Whereas at D_1 receptor sites dopamine acts at
micromolar concentrations and butyrophenone neuroleptics are
weak antagonists (μM), at D_2 receptor sites dopamine is more
potent and butyrophenone neuroleptics are active in the
nanomolar range. These latter pharmacological properties are
observed in the few systems where dopamine has been found to
inhibit adenylate cyclase. However, a more detailed and
complex classification containing up to four different subtypes
(D_1-D_4) has been recently proposed ($\underline{6,7}$). This
classification which is based on differential labelling of
sites with various agonist and antagonist ligands as well as
pharmacological and other differences between these sites has
been reviewed by Seeman ($\underline{8}$).
       In the anterior pituitary gland where dopamine regulates
the secretion of prolactin release we identified dopaminergic
binding sites with the specificity of the D_2 subtype of
receptor by direct ligand binding of the ergot alkaloid
[^{3}H]dihydroergocryptine ($\underline{9}$). Using this technique, it was
possible for the first time to correlate the pharmacological
properties of these sites with a direct physiological effect,
the inhibition of prolactin release ($\underline{9}$). More recently, it has
been possible to correlate, in the pituitary gland, occupancy

of the receptor with inhibition of the enzyme adenylate cyclase activity (10) or cyclic AMP levels (11). In the brain where these sites had been identified previously using neuroleptics no such correlation had been possible.

Because of the complex relationship between the binding of the numerous agonist and antagonist ligands for the dopamine receptor and the differences in characteristics and numbers of the sites labelled by these ligands, we set out to examine the properties of the dopamine receptor of the porcine anterior pituitary gland using both an agonist radioligand, [^{3}H]n-propylapomorphine, and an antagonist, [^{3}H]spiroperidol. Quantitative analysis of these data suggest that both ligands interact with a single dopamine receptor in the anterior pituitary gland. This receptor, however, appears to exist in two distinct affinity forms reciprocally favored by agonists and antagonists. One of the two forms of the receptor, the agonist high affinity/antagonist low affinity form may result from the stable association of the receptor with a guanine nucleotide binding protein. Furthermore, this form of the receptor can be solubilized with digitonin; however only upon prelabelling of the receptor site with the labelled agonist can nucleotide-sensitivity be returned. This paper will summarize the evidence supporting the formulation that a single dopamine receptor labelled by [^{3}H]spiroperidol exists in two distinct affinity forms in porcine anterior pituitary gland (12,13). This formulation appears as an attractive alternative to the proposal of Seeman and coworkers (8) of two distinct subtypes designated D_2 and D_4 for this receptor.

Quantitative Relationship of Agonist and Antagonist Ligand Binding to Porcine Anterior Pituitary Gland Membranes

In particulate preparations of porcine anterior pituitary gland prepared by differential centrifugation (12), the agonist [^{3}H]n-propylapomorphine and the antagonist [^{3}H]spiroperidol bind with the same appropriate dopaminergic specificity (Figure 1A and B) (i.e. n-propylapomorphine$_3$> apomorphine > ADTN > dopamine) (ADTN not shown for the [^{3}H]NPA curves). It should be noted that the agonist competition curves for [^{3}H]spiroperidol binding are complex or multiphasic with slope factors of less than 1. Quantitative analysis of these data by linear least square fitting (14) indicate that the agonists compete for the sites labelled by [^{3}H]spiroperidol with two different affinities; roughly 50% of the total [^{3}H]spiroperidol sites exhibit high affinity for the agonists and the remaining sites have lower affinity for the same agonists. The ratio of the two agonist affinities (K_H/K_L) vary from 30 in the case of apomorphine ($K_H = 4.7 \pm 2.4$ hM,

K_L = 160 ± 61 nM) to 200 for n-propylapomorphine (K_H = 0.081 ± 0.022 nM, K_L = 18 ± 3 nM). On the other hand, agonist competition curves for [^{3}H]n-propylapomorphine are monophasic with slope factors around unity (Figure 1B) suggesting that both the labelled agonist and the competing agonist interact with a single form of the receptor. The K_D values calculated from the agonist competition of [^{3}H]n-propylapomorphine correspond closely to those obtained from the higher of the two affinity forms evidenced in the agonist competition curves of [^{3}H]spiroperidol (e.g. for n-propylapomorphine: K_H ([^{3}H]spiroperidol) = 0.081 ± 0.022 nM vs. K_D ([^{3}H]n-propylapomorphine) = 0.075 ± 0.022 nM).

These data suggest that whereas [^{3}H]spiroperidol labels the entire population of the dopamine receptor, only a portion of these same sites are labelled with [^{3}H]n-propylapomorphine. This is confirmed by findings from saturation isotherms which show that in a given membrane preparation [^{3}H]n-propylapomorphine labels with high affinity (K_D = 0.26 nM) only one-half as many sites as [^{3}H]spiroperidol. The remaining [^{3}H]spiroperidol sites possess affinity that is too low for agonists (K_D = 18-20 nM for n-propylapomorphine) to be labelled by direct binding with [^{3}H]n-propylapomorphine with the concentrations of ligand normally used to perform saturation isotherms (10 pM to 1 nM). Therefore, it appears that by direct binding, [^{3}H]n-propylapomorphine labels only the agonist high affinity form of the receptor population labelled by [^{3}H]spiroperidol.

Guanine Nucleotides Modulation of Receptor Affinity for Agonists

Guanine nucleotides have been implicated in several receptor systems as important modulators of the affinity of receptors for their specific hormones or agonists (15). The effects are usually observed as a decrease in the affinity of the receptor for agonists (16). For example, in the beta-adrenergic receptor system, guanine nucleotides have been shown to reduce the ability of agonists to compete for antagonist binding to the receptor (17). In the beta-adrenergic system, agonists promote the formation of a ternary complex (18) composed of the hormone, receptor and a nucleotide binding protein (19). This ternary complex is destabilized or its formation inhibited in the presence of guanine nucleotides (20). In dopaminergic receptor systems, similar effects of guanine nucleotides on agonist potency at the receptor have been reported (reviewed in (8) and c.f. refs. (19-24) cited in (12)). In order to probe the relationship of the two different affinity forms of the dopamine receptor, it was of interest to

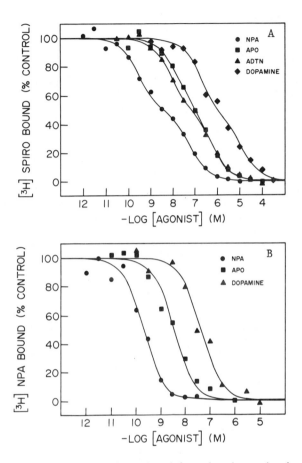

Figure 1. Competition curves of a series of dopaminergic agonists for the direct binding of [³H] agonist and antagonist ligands.

Porcine anterior pituitary gland membranes prepared as described previously (12), were incubated with [³H]spiroperidol (~ 140 pM) (A) or [³H]n-propylapomorphine (~ 150 pM) (B) and increasing concentrations of the various agonists as shown. Membranes suspended in 50 mM Tris-HCl, 6 mM MgCl₂, 1 mM EDTA, 100 mM NaCl, 0.1% ascorbate, 10 µM pargyline, pH 7.4, at 25 °C were incubated for 60 min at 25 °C in a total volume of 1 mL containing 0.3 mg and 0.6 mg of membrane protein/assay for [³H]spiroperidol and [³H]n-propylapomorphine, respectively. Binding was initiated by the addition of membranes, the incubations were terminated by adding 5 mL of cold 25 mM Tris-HCl, 2 mM MgCl₂ (pH 7.4) (4 °C) and rapid vacuum filtration on GF/C or GF/B glass fiber filters with four additional 5 mL washings. Each agonist drug competed for binding to the level of nonspecific binding that was determined in the presence of 1 µM (+)butaclamol. Under routine conditions specific binding accounted for 80–90% of total binding for [³H]spiroperidol and 50–70% for [³H]n-propylapomorphine. In the experiment shown, 100% [³H]spiroperidol binding corresponded to 32 pM or ~ 107 fmol/mg whereas 100% [³H]n-propylapomorphine was 33 pM or 55 fmol/mg. The points represent experimental data, and the lines represent the best fit of the data according to a model for one or two binding sites analyzed as described previously (12). The experiment shown was performed in duplicates and is representative of three (A) and two (B) such experiments. (Reproduced with permission from Ref. 12. Copyright 1982, American Society for Pharmacology and Experimental Therapeutics.)

examine the effect of guanine nucleotides on dopaminergic
ligand binding to porcine anterior pituitary membranes.

As shown in Figure 2, the guanine nucleotide Gpp(NH)p
markedly decreases the apparent potency of an agonist such as
n-propylapomorphine in competing for the binding of the
antagonist [^{3}H]spiroperidol. The control competition curve
which is biphasic and shallow is shifted to lower apparent
potency and steepened in the presence of guanine nucleotides.
In addition, as will be discussed below, an apparent increase
in [^{3}H]spiroperidol binding was observed in the presence of
guanine nucleotides. Quantitative analysis of the control
competition curve indicated that 50% of the [^{3}H]spiroperidol
sites display high affinity for the agonist (K_H = 0.29 ± 0.13
nM) and 50% of the sites displayed a lower affinity (K_L = 8.3
± 2.0 nM). In the presence of nucleotide, however, the
proportion of the agonist high affinity form is reduced to only
10 to 15% with no significant change in affinity (K_H = 0.20 ±
0.35 nM). The remaining 85 to 90% of the sites, then show low
affinity for the agonist indistinguishable from the lower
agonist affinity of the control curve (8.3 ± 2.0 nM vs. 12.0 ±
1.9 nM). Thus, guanine nucleotides appear to decrease the
proportion of the receptor exhibiting high affinity for
agonists (12). This effect can be tested directly by examining
the effects of nucleotides on direct agonist binding. As shown
in Figure 3 in the presence of increasing concentrations of
GTP, there is a progressive loss of the number of agonist high
affinity form of the receptor detectable by direct [^{3}H]n-
propylapomorphine binding (control = 26.7 ± 2.1 pM; + 10 μM GTP
= 19.6 ± 2.4 pM and + 1 mM GTP = 6.0$_3$± 1.4 pM). The affinity
of the remaining sites labelled by [^{3}H]n-propylapomorphine
remains essentially the same for the ligand.

The effect of guanine nucleotides on this system, much
like in other receptor systems (18), is to decrease the
proportion of the receptor existing in an agonist high affinity
form. This effect appears as either an apparent decrease in
the ability of agonists to compete for antagonist radioligand
binding or an actual decrease in the number of sites labelled
with high affinity by a labelled agonist. This latter effect
is due to the fact that the agonist lower affinity form of the
receptor usually cannot be labelled by direct agonist ligand
binding. Thus, these results are consistent with the thesis
that the dopamine receptor in the porcine anterior pituitary
gland can exist in two different affinity forms with respect to
agonists and that these two states can be interconverted by
guanine nucleotides. These specific effects of guanine
nucleotides strongly suggest that high affinity interactions of
the receptor with agonists involve another component which
mediates the effects of nucleotides.

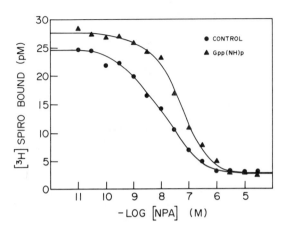

Figure 2. Effect of the guanine nucleotide Gpp(NH)p on competition of the agonist n-propylapomorphine for [³H]spiroperidol binding.

Membranes were incubated as described in Figure 1 with [³H]spiroperidol (~ 160 pM) and increasing concentrations of n-propylapomorphine in the presence and absence of 100 µM Gpp(NH)p. The lines are the best fit obtained with a model for two affinity forms of the receptor as described previously (14). The experiment shown is representative of 10–12 experiments. (Reproduced with permission from Ref. 12. Copyright 1982, American Society for Pharmacology and Experimental Therapeutics.)

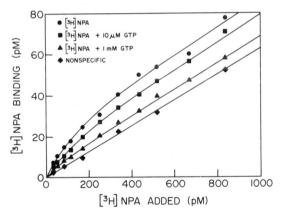

Figure 3. Effects of guanine nucleotides on the direct binding of [³H]n-propyl-apomorphine.

Porcine anterior pituitary membranes were incubated with increasing concentrations (10 pM–900 pM) of [³H]n-propylapomorphine in the presence of 10 µM and 1 mM GTP and in the absence of GTP as the control. Nonspecific binding was determined in the presence of 1 µM (+)butaclamol and was the same with or without the addition of GTP. Data were analyzed as described in Figure 1. The number of sites labelled by [³H]n-propylapomorphine was 26.7 pM (absence), 19.6 pM in the presence of 10 µM GTP and 6.0 pM in the presence of 1 mM GTP and dissociation constants for the ligand were 280 and 210 pM. The experiment was performed in duplicate and is representative of three such experiments.

Guanine Nucleotides Modulation of Receptor Affinity for Antagonists

As can be observed from competition experiments of n-propylapomorphine for [^{3}H]spiroperidol binding (Figure 2) the presence of guanine nucleotides yields an apparent increase in the binding of [^{3}H]spiroperidol. This effect is quite different from the usual effect of nucleotides observed on agonist binding (15,18). However, effects of guanine nucleotides on antagonist binding have been described for muscarinic receptor systems (22,23) and more recently for the beta-adrenergic receptor system (24). In order to explore further this effect detailed saturation binding isotherms of [^{3}H]spiroperidol binding were performed in the presence and absence of a guanine nucleotide. As shown in Figure 4, the control saturation curve as represented in the Scatchard plot coordinates is slightly curvilinear but becomes linear in the presence of 1 mM GTP. Quantitative analysis by linear least square fitting indicates that two K_D values (45 and 415 pM) can be significantly derived for [^{3}H]spiroperidol binding in the absence of GTP and a single value of 56 pM (not significantly different than 45 pM) in the presence of GTP. The two affinity forms which are present in roughly equal proportions in the control curve are converted to a single high affinity form in the presence of GTP with no change in the total number of sites for [^{3}H]spiroperidol (25). GTP displays the same dose response (EC$_{50}$ = 32 μM) for its effect on [^{3}H]spiroperidol binding as the effect on agonist binding (EC$_{50}$ = 17 μM). Thus, it appears that the same two affinity forms of the receptor which are discriminated by agonists may also be discriminated by antagonists although in a reciprocal fashion. The discriminating power of [^{3}H]spiroperidol for the two forms of the receptor is, however, much weaker than for agonists (2-10 fold for [^{3}H]spiroperidol and 30-200 fold for agonists). Thus, it would appear that in the presence of nucleotides the agonist high affinity form/antagonist low affinity form is converted to the agonist low affinity/ antagonist high affinity form of the receptor. The possible physiological relevance of these findings will be discussed later in this chapter.

N-ethylmaleimide and Heat Treatments of Membranes Mimic the Effect of Guanine Nucleotides on Ligand Binding

All the data presented above point to the involvement of a putative guanine nucleotide binding protein in the interaction of agonists with the dopamine receptor in the porcine anterior pituitary gland. We, therefore, wanted to obtain more compelling evidence for the interaction of a nucleotide binding

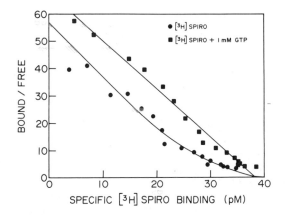

Figure 4. Effects of GTP on the binding isotherms of the antagonist [³H]spiro-peridol.

Porcine anterior pituitary membranes were incubated as described in Figure 1 with [³H]spiro-peridol (10–1100 pM). Nonspecific binding was determined in the presence of 1 μM (+)buta-clamol. Data were analyzed by curve fitting with a model for the binding of the radioligand to one or two classes of sites (14). Data are presented in the Scatchard plot coordinates. The experiment was performed in duplicate and is similar to several other experiments with the same results but it represents an extreme case of the ability of the [³H] antagonist to discriminate between the two forms of the receptor. The two dissociation constants calculated for the control curve are different by about 10-fold. (Reproduced with permission from Ref. 25. Copyright 1982, American Society for Pharmacology and Experimental Therapeutics.)

entity with the receptor. N-ethylmaleimide (NEM), a sulhydryl alkylating agent, mimics the effects of nucleotides on agonist binding at the well-characterized beta-adrenergic receptor of the frog erythrocyte (26). In a manner similar to the effect of guanine nucleotides in that system, NEM prevents the formation of the agonist high affinity form of receptor with no change in total number of receptors. Moreover, it is apparent that the guanine nucleotide regulatory protein activity of the beta-adrenergic receptor demonstrates exquisite sensitivity to temperature. Ross et al. (27) showed that incubation of S-49 lymphoma cell membranes at 50° inactivates the regulatory protein activity with a $t\frac{1}{2}$ of 2-8 minutes as assayed by reconstitution studies. The regulatory protein activity assayed by reconstitution is the same functional entity which interacts with the beta-adrenergic receptor to form the high affinity agonist ternary complex (19). Therefore, it is reasonable to assume that NEM and heat treatment interfere with the receptor-affinity modulating function of the guanine nucleotide regulatory protein in that system.

Treatment of porcine anterior pituitary gland membranes with either NEM at concentrations between 1-100 μM or heat (53° for 4 min.) qualitatively and quantitatively mimics the effects of guanine nucleotides on dopaminergic ligand binding. As shown in Figures 5A and B, competition curves of the agonist n-propylapomorphine for [^{3}H]spiroperidol binding are shifted to the right and steepened. These effects are identical to the ability of guanine nucleotides to decrease the proportion of agonist high affinity form of the receptor. Similarly, increasing concentrations of NEM from 1 μM to 100 μM cause a progressive decrease in the direct binding of the agonist [^{3}H]n-propylapomorphine (13). Moreover, it can be observed that both NEM and heat treatments seem to affect [^{3}H]spiroperidol binding in a manner similar to guanine nucleotides since an apparent increase in binding of [^{3}H]spiroperidol can be observed (Figures 5A and B). Heat treatment of membranes does not increase [^{3}H]spiroperidol binding as much as GTP because under these conditions (53° for 4 min.) [^{3}H]spiroperidol binding is slightly inhibited (15-20%) (not shown) (13).

Thus, these results suggest that the effects of heat and NEM treatments mimic the modulation of dopaminergic ligand binding by guanine nucleotides and are very similar to their effects on beta-adrenergic systems. The only exception appears to be the relative sensitivity of the dopaminergic system to NEM (13,26). Whereas, on dopaminergic binding, NEM elicits its effects with an EC_{50} = 6 μM, concentration in the millimolar range are required for the same effects on beta-adrenergic agonist binding. By analogy with the beta-adrenergic receptor system, these data support an interaction of a guanine

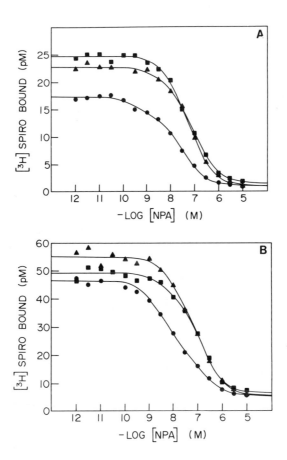

Figure 5. Comparison of the effects of NEM and heat treatments and the presence of GTP on the ability of n-propylapomorphine (agonist) to compete for the binding of the antagonist [³H]spiroperidol.

Membranes were incubated with [³H]spiroperidol (∼ 109 pM) for 1 h at 25 °C and increasing concentrations of n-propylapomorphine (A). Key to A: ●, *control membranes;* ■, *membranes treated in the presence of 100 μM NEM as described below; and* ▲, *membranes incubated in the presence of 1 mM GTP. The curves represent computer drawn lines that best fit the data points. NEM treatment was performed as follows: Membranes were incubated with the indicated concentration of NEM for 30 min at 25 °C prior to addition of the radioligand. The same results were obtained whether or not the membranes were washed free of residual NEM prior to the binding assay. All three curves are from a single experiment that is representative of three experiments. Control membranes and membranes treated at 53 °C for 4 min as described below were incubated with [³H]spiroperidol (190 pM) in the presence of increasing concentrations of n-propylapomorphine. One set of membranes were also incubated in the presence of 100 μM GTP. Data were analyzed by computer-based methods as described in Figure 1 with the lines representing the best fit to the data (14). The experiment was performed and is representative of two such experiments. (Reproduced with permission from Ref. 13. Copyright 1982, American Society for Pharmacology and Experimental Therapeutics.)*

nucleotide regulatory protein with the dopamine receptor in the anterior pituitary gland.

Solubilization of the Dopamine Receptor

One of the prerequisites for the purification and characterization of a membrane-bound hormone receptor is its solubilization in an active form, i.e. with retention of the ability of the receptor to interact with specific ligands. Several reports have appeared already documenting the solubilization of dopaminergic binding sites from brain tissue (28-31) using detergents such as digitonin and 3'-[(3 cholamidopropyldimethylamino)dimethyl-amonio]-1-propanesulfonate (CHAPS) or chaotropic agents. In the porcine anterior pituitary gland we report here that the dopamine receptor can be solubilized using the detergent digitonin.

Treatment of membrane preparations with digitonin (1-2%) in 100 mM NaCl, 10 mM Tris-HCl pH 7.4 releases 20-30% of the membrane-binding sites in a soluble form that cannot be sedimented at 100,000 xg for 60 min. Saturation isotherms of [^{3}H]spiroperidol as measured by a Sephadex G-50 assay (32) for separation of bound from free ligand reveal a dissociation constant (K_D) of 0.5 nM for spiroperidol (not shown). Although, this value is somewhat lower than the K_D of 40-70 pM obtained for [^{3}H]spiroperidol binding to particulate preparations, the specificity of binding of the antagonist to the solubilized sites is similar to that in the membrane fraction (n-propylapomorphine > ADTN > dopamine) (not shown) (12). As mentioned previously, agonist competition curves for [^{3}H]spiroperidol binding in particulate preparations are biphasic and high and low affinity forms of the receptor for agonists can be documented; in solubilized preparations a single low affinity can be detected for the competition of agonists for soluble [^{3}H]spiroperidol binding. By comparison to the K_D values (0.080 and 18 nM) calculated for n-propylapomorphine competition of particulate [^{3}H]spiroperidol binding, the value obtained from solubilized preparations is 35 nM. Moreover, the presence of guanine nucleotides in the assays does not further reduce the ability of the agonists to compete for [^{3}H]spiroperidol binding (not shown). Therefore, it appears that solubilization interferes with the ability of agonists to interact with high affinity with the receptor. This is a common finding in several solubilized receptor systems (33).

However, if the agonist high affinity form of the receptor is labelled prior to solubilization by incubation of membranes with [^{3}H]n-propylapomorphine then the agonist high affinity form of the receptor appears to be stable to solubilization (Table I). Shown in Table I are the results of three

Table I

Solubilization by Digitonin of Agonist and Antagonist
Prelabelled Dopamine Receptor from Anterior
Pituitary Gland Membranes

$[^3H]$ Ligand

Experiment	$[^3H]$NPA	$[^3H]$Spiro	$[^3H]$Spiro +1 nM NPA	% agonist high affinity sites
	fmol	fmol	fmol	%
1	63	155	110	41
2	55	133	90	42
3	35.9	60	--	59

Membranes were incubated for 60 minutes at 25° as described in
the legend to Figure 1 with 800 pM $[^3H]$n-propylapomorphine
(NPA) and 700 pM $[^3H]$spiroperidol (Spiro) and $[^3H]$spiroperidol
in the presence of 1 nM unlabelled NPA, a concentration
sufficient to saturate the agonist high affinity form of the
receptor. After the incubation, membranes were centrifuged and
solubilized by resuspending the membranes in 1% digitonin, 100
mM NaCl, 25 mM Tris-HCl 2 mM $MgCl_2$, 0.1% ascorbate pH 7.4 at
4°C, stirring on ice for 30 minutes and centrifuging at 40,000
x g for 45 minutes. Specific binding for both ligands was
determined by Sephadex G-50 chromatography of the solubilized
samples from incubation with ligands alone or ligands with
10^{-5}M (+)butaclamol. Results are expressed in fmol of
specific binding in 1 ml of solubilized preparation in the
respective experiments. % agonist high affinity sites is
expressed as the percent of $[^3H]$NPA sites compared to the
number of sites labelled with $[^3H]$spiroperidol.

experiments in which membrane preparations were incubated with [^{3}H]n-propylapomorphine, [^{3}H]spiroperidol and [^{3}H]spiroperidol in the presence of 1 nM unlabelled n-propylapomorphine prior to solubilization with digitonin. Under these conditions, the results indicate that about the same proportion of agonist high affinity form of the receptor can be solubilized as that which can be evidenced in membrane preparations (i.e. about 50%). In the presence of 1 nM n-propylapomorphine, a concentration sufficient to saturate the agonist high affinity form of the receptor in the membrane, the amount of [^{3}H]spiroperidol binding solubilized is reduced by approximately the amount of [^{3}H]n-propylapomorphine sites solubilized. In addition, the data in Table II indicate that the solublized agonist high affinity form of the receptor remains sensitive to the effect of nucleotides. [^{3}H]n-Propylapomorphine binding, assayed in the membranes in the presence of Gpp(NH)p or treated with Gpp(NH)p after solubilization, is reduced to about the same extent. As shown in Figures 6A and B, this apparent decrease in the number of binding sites for [^{3}H]n-propylapomorphine in the presence of guanine nucleotides in both membranes (Figure 6A) and solubilized preparations (Figure 6B), can be attributed to a decrease in the affinity of the receptor for the agonist. This apparent decrease in affinity can be accounted by the increased dissociation rate of [^{3}H]n-propylapomorphine in both preparations in the presence of Gpp(NH)p.

The results presented above are consistent with the formulation postulated previously, that the receptor can exist as two affinity forms, one of which is highly prefered or stabilized in the presence of an agonist. In addition, these data indicate that whereas the receptor can be solubilized with apparent retention of its binding specificity, high affinity interactions of the receptor with agonists are not present in solubilized preparations unless stabilized prior to solubilization by the presence of an agonist. These results are entirely analogous to the beta-adrenergic receptor system (34) where the component required for high affinity agonist interactions with the receptor has been documented to be a guanine nucleotide binding protein.

Conclusions

The discrimination in the membrane preparations by agonists of two affinity forms of the receptor allows the comparison of the potency of these agonists with their ability to elicit inhibition prolactin release. It can be observed, that the K_D values of agonists observed for the high affinity form of the receptor (K_H) correlate very closely with the ability (EC_{50}) of these same agents to inhibit prolactin release (9) (K_H (dopamine) = 66 nM vs. EC_{50} (dopamine) = 35

Table II

Effect of Guanine Nucleotides on High Affinity Agonist Binding
to Membrane and Solubilized Receptor Preparations

	[³H]NPA specific binding (fmol)			
	Membrane preparations		Solubilized preparations	
Experiment	Control	+100µMGpp(NH)p	Control	+100µMGpp(NH)p
1	193		40	8.6 (79)
2	132	19 (86)	44	13.0 (71)

Membranes were incubated as described in Table I with 800 pM [³H]n-propylapomorphine (control) and in the presence of 100 µM Gpp(NH)p. Specific binding was determined as described in the legend to Figure 1. For data with solubilized preparations, membranes were incubated with 800 pM [³H]n-propylapomorphine for 60 minutes at 25°C then solubilized as described in Table I. The solubilized samples were then incubated for 15 hours at 4° in the presence or absence of 100 µM Gpp(NH)p and specific binding determined by Sephadex G-50 chromatography from the difference between the above samples and samples which had been incubated with 10^{-5}M (+)butaclamol from the initial incubation. The numbers in parenthesis indicate the percent decrease in agonist binding in the presence of Gpp(NH)p.

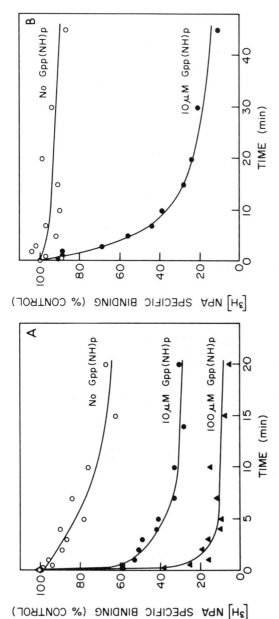

Figure 6. Effects of guanine nucleotides on the dissociation rate of [³H]n-propylapomorphine binding from membrane-bound and solubilized receptor preparations.

Membranes were incubated as described in Figure 1 with 500 pM [³H]n-propylapomorphine for 60 min at 25 °C to achieve equilibrium in the presence of 0.1% ascorbate (A). Dissociation was initiated by adding at time 0 an excess of (+)butaclamol (10⁻⁵ M) or (+)butaclamol, plus 10 and 100 µM Gpp(NH)p. Samples were filtered on GF/B filters at time intervals shown. Nonspecific binding was determined by parallel incubations containing (+)butaclamol from the beginning of the incubation. The experiment performed in duplicate is representative of three such experiments.   Membranes were also incubated as above with [³H]n-propylapomorphine to achieve equilibrium (B). After the incubation, membranes were centrifuged and resuspended in 1% digitonin 25 mM Tris-HCl, 100 mM NaCl, 2 mM MgCl₂, 0.1% ascorbate, pH 7.4, at 4 °C, stirred for 30 min on ice, and centrifuged at 40,000 × g for 45 min to obtain the solubilized labelled receptor preparation. Dissociation of the ligand was started by adding at time 0 excess (+)butaclamol (10⁻⁵ M) or (+)butaclamol and 100 µM Gpp(NH)p. Samples were incubated at 4 °C for the time intervals indicated, and binding was determined by Sephadex G-50 chromatography. The results are representative of two similar experiments.

nM, $K_{H(apomorphine)}$ = 4.7 nM vs. $EC_{50}(apomorphine)$ = 3 nM). These correlations are much closer than when the overall relative potency of an agonist for competing for ligand binding is compared with the physiological response (9). It is therefore reasonable to postulate that the physiological response to dopamine, i.e. inhibition of prolactin release, is a reflection of the high affinity interaction of dopaminergic agonists or dopamine itself with the receptor in the anterior pituitary gland.

The discrimination of the same two forms of the receptor by antagonists in a reciprocal fashion to agonists is much more subtle. First, the differences in affinity observed for the antagonist [³H]spiroperidol are small (2-10 fold) as compared to agonits (30-200 fold), hence more difficult to demonstrate unless extremely detailed saturation isotherms are performed. Moreover, the discrimination of the two affinity forms of the receptor by an antagonist ligand has been demonstrated for [³H]spiroperidol only as no other antagonists have been examined. The fact that antagonists in several receptor systems can slightly distinguish two forms of a receptor probably has no major physiological significance with respect to the actions of antagonists. Nonetheless, the quantitative aspect of the binding of [³H]spiroperidol to the two forms of the receptor and their modulation by guanine nucleotides can be taken as further support for the existence of these two different interconvertible forms of the receptor which appear more important in the mechanism of action of agonists.

In the beta adrenergic receptor system of the frog erythrocyte (18) the amount of agonist high affinity state (R_H) of the receptor induced by an agonist and the ratio of K_L/K_H were found to correlate closely wtih the intrinsic activity of the agonist in stimulating adenylate cyclase. In the alpha$_2$-adrenergic receptor system of the human platelets, R_H did not correlate with the intrinsic activity of agonists for their ability to attenuate adenylate cyclase stimulation (36). In that system, only the ratio of K_L/K_H appeared to correlate with the intrinsic activity of the alpha-adrenergic agonists. Here, in anterior pituitary membranes constant proportions of both receptor states were evidenced but ratios of 30-200 were found for K_L/K_H for a series of dopaminergic agonists despite the fact that these agonists all appear to display full intrinsic activity in their ability to inhibit prolactin secretion from pituitary cells.

In addition to the effects of guanine nucleotides on the direct binding of agonists and antagonists, several factors point toward the interaction of a putative nucleotide binding protein in the mechanism of ligand binding to the dopamine receptor of anterior pituitary gland. First, both heat and NEM treatment of membranes mimic the qualitative and quantitative

effects of nucleotides on the binding of agonists and
antagonists. These treatments have been shown in other systems
to inactivate guanine nucleotide regulatory protein function.
Second, upon solubilization the ability of the receptor to
interact with high affinity with agonists is lost. However,
labelling of the receptor in membranes with the agonist
[^{3}H]n-propylapomorphine prior to solubilization produces a
stable agonist-receptor complex sensitive to guanine
nucleotides. These data strongly suggest the interaction of a
guanine nucleotide regulatory protein in the formation or
stabilization of the agonist high affinity form of the
receptor.

     We envisage that the formulation proposed here of the same
receptor site existing in two distinct affinity forms could
possibly represent the same entity as the D_2 and D_4 subtypes
of receptors proposed by Seeman (8). However, much more work
on the biochemical characterization of the receptors from both
pituitary gland and the brain will be necessary before a firm
conclusion can be reached.

     The dopamine receptor in the pituitary gland has the
pharmacological characteristics of the D_2 subtype of receptor.
Recently, dopamine has been found to decrease cyclic AMP
accumulation or inhibit adenylate cyclase in the intermediate
lobe and the anterior pituitary gland (2,3,4,10,11,35). Thus,
this dopamine receptor may be linked physiologically to an
inhibition of adenylate cyclase similarly to the alpha$_2$-
adrenergic receptor. Therefore, the high affinity interaction
of the receptor with agonists and its modulation by guanine
nucleotides must be implicated in the biochemical translation
of agonist occupancy of the receptor to the effector system,
adenylate cyclase. It has been suggested (15) that a putative
nucleotide binding protein different from that implicated in
hormone stimulation of adenylate cyclase may be involved with
attenuating systems. Further studies will be required,
however, in an attempt to demonstrate a direct physical interaction of the receptor with such a guanine nucleotide binding
protein and its relationship to that implicated in stimulatory
systems.

Literature Cited
 1. Kebabian, J.W.; Petzold, G.L.; Greengard, P. Proc. Natl.
     Acad. Sci. USA 1972, 69, 2145-2149.
 2. Cote, T.E.; Grewe, G.W.; Kebabian, J.W. Endocrinology
     1982, 110, 805-811.
 3. Meunier, H.; Labrie, F. Life Sci. 1982, 30, 963-968.
 4. Giannattasio, G.; DeFerrari, M.E.; Spada, A. Life Sci.
     1981, 28, 1605-1612.

5. Kebabian, J.W.; Calne, D.B. Nature 1979, 277, 93-96.
6. Titeler, M.; List, S.; Seeman, P. Psychopharmacology 1979, 3, 411-420.
7. Titeler, M.; Seeman, P. Eur. J. Pharmacol. 1979, 56, 291-292.
8. Seeman, P. Pharmacol. Rev. 1980, 32, 229-313.
9. Caron, M.G.; Beaulieu, M.; Raymond, J.; Gagne, B.; Drouin, J.; Lefkowitz, R.J.; Labrie, F. J. Biol. Chem. 1978, 253, 2244-2253.
10. Munenura, M.; Cote, T.E.; Tsuruta, K.; Eskay, R.L.; Kebabian, J.W. Endocrinology 1980, 106, 1676-1683.
11. Labrie, F.; Borgeat, T.; Barden, N.; Godbout, M.; Beaulieu, M.; Ferland, L. "Polypeptide Hormones" (Beers, R.J., Jr. and Bassett, E.G., eds.); Raven Press: New York, 1980; pp. 235-251.
12. De Lean, A.; Kilpatrick, B.F.; Caron, M.G. Mol. Pharmacol. 1982, 22, 290-297.
13. Kilpatrick, B.F.; De Lean, A.; Caron, M.G. Mol. Pharmacol. 1982, 22, 298-303.
14. De Lean, A.; Hancock, A.A.; Lefkowitz, R.J. Mol. Pharmacol. 1982, 21, 5-16.
15. Rodbell, M. Nature 1980, 284, 17-22.
16. Rodbell, M.; Krans, H.M.J.; Pohl, S.L.; Birnbaumer, L. J. Biol. Chem. 1971, 246, 1872-1876.
17. Lefkowitz, R.J.; Mullikin, D.; Caron, M.G. J. Biol. Chem. 1976, 251, 4686-4692.
18. De Lean, A.; Stadel, J.M.; Lefkowitz, R.J. J. Biol. Chem. 1980, 255, 7108-7117.
19. Stadel, J.M.; Shorr, R.G.L.; Limbird, L.E.; Lefkowitz, R.J. J. Biol. Chem. 1981, 256, 8718-8723.
20. Kent, R.S.; De Lean, A.; Lefkowitz, R.J. Mol. Pharmacol. 1980, 17, 14-23.
21. Stadel, J.M.; De Lean, A.; Lefkowitz, R.J. Adv. Enzymol. 1982, 53, 1-43.
22. Ehlert, F.J.; Roeske, W.R.; Yamamura, H.I. J. Supramol. Struct. 1980, 14, 149-155.
23. Burgisser, E.; De Lean, A.; Lefkowitz, R.J. Proc. Natl. Acad. Sci. USA 1982, 79, 1732-1736.
24. Wolfe, B.B.; Harden, T.K. J. Cyclic Nucleotide Res. 1981, 7, 303-312.
25. De Lean, A.; Kilpatrick, B.F.; Caron, M.G. Endocrinology 1982, 1064-1066.
26. Stadel, J.M.; Lefkowitz, R.J. Mol. Pharmacol. 1979, 16, 709-718.
27. Ross, E.M.; Howlett, A.C.; Ferguson, K.M.; Gilman, A.G. J. Biol. Chem. 1978, 253, 6406-6412.
28. Gorissen, H.; Laduron, P. Nature 1979, 279, 72-74.
29. Davis, A.; Madras, B.K.; Seeman, P. Eur. J. Pharmacol. 1981, 70, 321-323.

30. Lew, J.Y.; Fong, J.C.; Goldstein, M.A. Eur. J. Pharmacol. 1981, 72, 403-405.
31. Clement-Cormier, Y.C.; Meyerson, L.R.; McIascc, A. Biochem. Pharmacol. 1980, 29, 209-216.
32. Caron, M.G.; Lefkowitz, R.J. J. Biol. Chem. 1976, 251, 2374-2384.
33. Caron, M.G.; Limbird, L.E.; Lefkowitz, R.J. Mol. Cell. Biochem. 1979, 28, 45-67.
34. Limbird, L.E. Biochem. J. 1981, 195, 1-13.
35. Onali, P.; Schwartz, J.P.; Costa, E. Proc. Natl. Acad. Sci. USA 1981, 78, 6531-6534.
36. Hoffman, B.B.; Michel, T.; Brenneman, T.B.; Lefkowitz, R.J. Endocrinology 1982, 110, 926-932.

RECEIVED January 7, 1983

Commentary: The Dopamine Receptor of the Anterior Pituitary Gland

SUSAN R. GEORGE, MASAYUKI WATANABE, and PHILIP SEEMAN

University of Toronto, Department of Pharmacology, Toronto, Canada

There is much evidence to suggest the presence of multiple dopamine receptor subtypes (1). Of these subtypes, the anterior pituitary (AP) contains a pure population of postsynaptic D_2 dopamine receptors, so that unlike studies in brain it is possible in AP to directly correlate receptor binding parameters to functional, physiological dopaminergic events such as the inhibition of prolactin secretion and the attenuation of adenylate cyclase activity. The D_2 receptor appears to exist in two states, one with high and the other lower affinity for dopamine agonists. Both forms appear to have high-affinity for dopamine antagonists. In general, agonist competition curves of ^{3}H-agonist binding are monophasic with high Hill coefficients (~1), and agonist competition curves of ^{3}H-antagonist binding are biphasic in AP with low Hill coefficients (<1). It is clear that it is impossible to delineate the individual components of such complex curves and derive quantitative parameters without detailed computer-assisted curve fitting, as has been so elegantly demonstrated by De Lean et al. for the β-adrenoceptor (2,3). Such quantitative analyses of radioligand binding studies yield important information regarding the dissociation constants and the relative proportions of each of the two sites while permitting statistical comparisons with other similarly generated curves.

The ternary complex model for the β-adrenoceptor has also been adopted for the D_2 dopamine receptor of the AP (4). The model suggests that the low-affinity state of the receptor, as detected by agonist binding, is a binary complex (agonist-receptor) and that the high-affinity state is a ternary complex (agonist-receptor-nucleotide binding protein), formed by the addition of a third membrane component [identified as the guanine nucleotide binding protein in other systems (5)]. Agonists bind with higher affinity to the ternary complex than to the binary complex and in addition, stabilize and promote the formation of the ternary complex. The coexistence of these binary and ternary complexes in the membrane provide a molecular basis for the characteristic biphasic, shallow curve of agonist

0097–6156/83/0224–0093$06.00/0

competition of antagonist binding. Antagonists are reported to
not discriminate between the binary and ternary complexes,
although there is some evidence to the contrary (6, Caron et al,
this book) to suggest that antagonists may do so, having higher
affinity for the binary complex than for the ternary complex.
More importantly, antagonists do not promote the formation of the
ternary complex, and their ability to discriminate between the
two forms of the receptor appears poor.

The results presented by Caron et al document that in the
porcine AP, the antagonist ligand ^{3}H-spiroperidol labels the
entire D_2 receptor population, with approximately 50% being in
the high-affinity state for agonists (H) and the remainder in
the lower affinity state (L). In keeping with this, the agonist
ligand ^{3}H-n-propylapomorphine in the concentration used (<1 nM)
labels only 50% as many sites as does ^{3}H-spiroperidol. Agonist
competition curves of ^{3}H-spiroperidol binding are complex and
resolved into two sites with differing dissociation constants K_H
and K_L, respectively for the high- and low-affinity states. The
agonists tested, n-propylapomorphine, apomorphine, ADTN and
dopamine all detect approximately 50% of receptors as being in
the high-affinity form. However, the ratio of the two dissociation constants (K_L/K_H) varies for each agonist and ranges from 30
for apomorphine to 200 for n-propylapomorphine.

These agonists all detect quantitatively equal proportions of
high- and low-affinity sites and this finding may be justified in
light of the fact that all these compounds are full dopaminergic
agonists and therefore can be presumed to have full intrinsic
activity. However, the wide range of values for K_L/K_H reported
for these full agonists (30-200) suggests that there may not be a
correlation between this parameter and intrinsic activity contrary to what has been reported for other systems (7,8).

In our system, also utilizing porcine AP, an example of
apomorphine competition of ^{3}H-spiroperidol is shown [Figure 1].
As indicated K_H = 3.40 ± 0.15 nM, K_L = 1300 ± 211 nM, with
K_L/K_H = 382. The fraction of sites exhibiting high affinity
for apomorphine is 33%, with 67% having lower affinity.

We also have preliminary evidence with several other dopaminergic agonists that the proportions of high and low affinity
sites varies with the agonists, in addition to differing K_L/K_H
ratios for each. It has previously been suggested that the
percentage of agonist high-affinity sites may be correlated with
the intrinsic activity of dopamine agonists (9) . The difference
in our observations may reflect differences in methodology, for
the absolute affinities K_H and K_L can vary greatly for any agonist
depending upon assay conditions.

The pivotal role of guanine nucleotides in mediating the
communication between the agonist-occupied receptor and the
effector adenylate cyclase, via the nucleotide binding protein
is established in several receptor systems. Guanine nucleotides
reduce the fraction of receptor present in the agonist high-

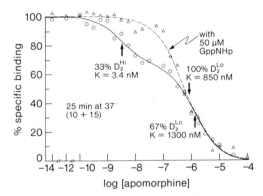

Figure 1. Apomorphine competition curves of [³H]spiroperidol binding, in the absence and presence of 50 µM GppNHp. Membranes were preincubated for 10 min at 37 °C and then incubated for 15 min at 37 °C with 100 pM [³H]spiroperidol. Specific binding was defined in the presence of 10 µM (−)-sulpiride, and accounted for 70–75% of total binding.

affinity state in a dose-dependent manner and convert them into
the low-affinity state. On a molecular basis, the likely explanation is that the ternary complex is destabilized or that its
formation is prevented by a decrease in the affinity of receptor
for the agonist (7). Agonist competition curves that were biphasic and shallow become steep as less and less of the high
affinity form is detected in the presence of guanine nucleotides.
It has been suggested in other systems that the extent of the
guanine nucleotide-induced shift in receptor affinity is directly
proportional to the intrinsic activity of the agonist in interacting with adenylate cyclase (10).

In porcine AP, as described in other receptor systems,
guanine nucleotides decrease the fraction of receptors that
can exist in the agonist high-affinity state. In the presence of
maximal effective concentrations of Gpp(NH)p, the nonhydrolyzable
analogue of GTP, the entire population of D_2 receptors labeled
by ^3H-spiperone exists in the low agonist affinity form, with no
detectable high-affinity form (Figure 1). This ability of
guanine nucleotides to convert all the receptors to the low
agonist affinity form is temperature-dependent. We find at 20°C
incubation temperatures that there is failure to convert a
fraction of the agonist high-affinity form of receptor (~40-50%
of R_H or 10-15% of R_{total}) that appears to be resistant or
insensitive to the effects of guanine nucleotides. However, at
37°C incubation temperature there is a consistent observation
that 100% of the receptors exist in the agonist low-affinity
form.

The total length of time of the 37°C incubation also appears
to be a factor in the proportion of receptors detected in the
agonist high-affinity state. Prolonged incubation at 37°C, before
there is any loss of total receptor numbers, leads to a progressive loss of the agonist high-affinity fraction (Figure 2)
so that eventually only the low agonist affinity form is seen to
exist (Figure 3). Unlike previous reports wherein brief exposure to high temperatures irreversibly reduced agonist high-
affinity binding (11,12), it is seen that increasing time of
incubation even at 37°C is cumulative and has similar effects.
This heat inactivation effect is identical to that seen in the
presence of guanine nucleotides, i.e. reducing agonist binding,
decreasing the proportion of the agonist high-affinity state of
the receptor and steepening agonist competition curves of antagonist binding. The effects of heat treatment and guanine nucleotides are not additive, implying a common mechanism. This
suggests therefore that the nucleotide binding protein is
thermolabile not just at high temperatures but even with prolonged exposure to 37°C.

It has recently become apparent that perhaps antagonists
can discriminate between the agonist high- and low-affinity
states in a reciprocal fashion to that of agonists (6, Caron et
al, this book). However, this ability to discriminate between

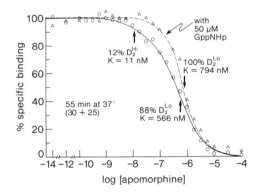

Figure 2. *Apomorphine competition curves of [³H]spiroperidol binding, in the absence and presence of 50 µM GppNHp. Membranes were preincubated for 30 min at 37 °C, then incubated for 25 min at 37 °C with 100 pM [³H]spiroperidol.*

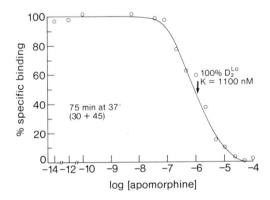

Figure 3. *Apomorphine competition curve of [³H]spiroperidol binding. Membranes were preincubated for 30 min at 37 °C, then incubated for 45 min at 37 °C with 100 pM [³H]spiroperidol.*

the two sites is considerably less than the ability of agonists
to do so (e.g. K_L/K_H = 9 for spiroperidol, 200 for n-propylapomorphine). Guanine nucleotides are reported to convert the
antagonist low-affinity form to the high-affinity form. Interestingly, quantitative analysis of antagonist binding reveals
that spiperone detects two sites present in approximately equal
proportions in the membrane, i.e. 50% of the receptors are in
the antagonist high-/agonist low-affinity state and 50% in the
antagonist low-/agonist high-affinity state. If this observation
is valid and correct, it creates a serious discrepancy in recon-
ciling the proposed ternary complex model with this analysis of
antagonist binding. Fundamental to this model is the concept
that the proportion of agonist high-affinity sites is dependent
upon the intrinsic activity of the agonist and that the agonist
stabilizes and promotes the generation of the ternary complex
(5,7,13). However, if the antagonist detects 50% of the receptors as existing in the agonist high-affinity form, then the
agonist high- and low-affinity forms must exist in the membrane
in such a ratio of equal proportions. If so, then the high-
affinity sites detected by agonists such as n-propylapomorphine
or apomorphine are not agonist-induced or agonist-specific and
therefore not a reflection of the intrinsic activity of the
agonist in question. The data thus far presented by Caron et al
can be explained as well by a model suggesting the presence of
two independent subclasses of receptors that do not interact in
the absence of guanine nucelotides.

In the β-adrenergic system, a similar estimate of 50%
agonist high affinity state has been described in the presence of
an antagonist (7,13), suggesting that spontaneously coupled
receptor-nucleotide binding protein can exist in the absence of
any agonist. However, in this system, β-adrenergic agonists
promoted the formation of as much as 92% agonist high-affinity
state of the receptor, contrary to that reported for the D_2
dopamine receptor of the AP. Thus, the ternary complex model
may have to be modified or an alternate sought to satisfactorily
explain all the observations with respect to dopaminergic agonist
and antagonist binding.

In any case, such direct binding measures of the dopamine
receptor in the AP provides new dimensions in the understanding
of the functional role of dopamine in the gland. The detailed
examination of the dynamics of the dopamine receptor in the AP
offers the simplest model of studying postsynaptic dopamine
receptors in isolation and will further our understanding of
dopamine receptor-transduced signals in the brain.

Literature Cited

1. Seeman, P. Pharmacol. Rev. 1980, 32, 229-313.
2. De Lean, A.; Munson, P.J.; Rodbard, D. Am. J. Physiol. 1978,
    235, E97-E102.

3. De Lean, A.; Hancock, A.A.; Lefkowitz, R.J. Mol. Pharm. 1982, 21, 5-16.
4. Sibley, D.R.; De Lean, A.; Creese, I. J. Biol. Chem. 1982, 257, 6351-6361.
5. Limbird, L.E.; Gill, D.M.; Lefkowitz, R.J. Proc. Natl. Acad. Sci. USA 1980, 77, 775-779.
6. De Lean, A.; Kilpatrick, B.F.; Caron, M.G. Endocrinology 1982, 110, 1064-1066.
7. Kent, R.S.; De Lean, A.; Lefkowitz, R.J. Mol. Pharm. 1980, 17, 14-23.
8. Hoffman, B.B.; Michel, T.; Brenneman, T.B.; Lefkowitz, R.J. Endocrinology 1982, 110, 926-932.
9. Sibley, D.R.; Leff, S.E.; Creese, I. Life Sci. 1982, 31, 637-645.
10. Lefkowitz, R.J.; Mulliken, D.; Caron, M.G. J. Biol. Chem. 1976, 251, 4686-4692.
11. Gurwitz, D.; Sokolovsky, M. Biochem. Biophys. Res. Commun. 1980, 94, 493-500.
12. Hamblin, M.W.; Creese, I. Mol. Pharm. 1982, 21, 52-56.
13. De Lean, A.; Stadel, J.M.; Lefkowitz, R.J. J. Biol. Chem. 1980, 255, 7108-7117.

RECEIVED November 29, 1982

4

Differentiation of Dopamine Receptors in the Periphery

LEON I. GOLDBERG and JAI D. KOHLI

The University of Chicago, Committee on Clinical Pharmacology, Departments of Pharmacological and Physiological Sciences and Medicine, Chicago, IL 60637

Two distinct dopamine (DA) receptors have been studied in the anesthetized dog. DA_1 receptors subserve smooth muscle relaxation and can be characterized by comparing putative DA agonists in the phenoxybenzamine pretreated renal arterial bed. DA_2 receptors located on sympathetic nerve endings subserve inhibition of norepinephrine release and can be characterized in the untreated femoral vascular bed by comparing vasodilation produced by dipropyl DA and putative agonists. The structural requirements for activation of DA_1 receptors are very strict. Positive compounds require 3,4-OH groups on the benzene ring. Substitutions on the alpha and beta carbon and single substitutions on the N atom, except for N-methyl, are inactive. N,N-di-substituted compounds with one group being N-propyl or N-butyl are 30-100 times less active than DA. For semi-rigid analogs related to dihydroxy-aminotetralins (ADTN), beta rotamers with 6,7-dihydroxy substitution showed DA_1 activity; the alpha rotamers with 5,6-dihydroxy-substitution were inactive except for N,N-di-n-propyl A-5,6-DTN, which was approximately equipotent to DA. 7,8-Dihydroxybenzazepines have been shown to be full or partial agonists. Apomorphine is a weak partial agonist; isoapomorphine and ergot derivatives are inactive. In contrast, many compounds are active on DA_2 receptors. These include compounds with 1 hydroxy group on the benzene ring and ergot compounds such as bromocriptine. Compounds active on both receptors exhibit differences in potency series. Apomorphine and dipropyl DA are much more potent as DA_2 agonists than as DA_1 agonists. Differences in antagonists

have also been demonstrated. Most compounds
studied antagonize both DA_1 and DA_2 receptors.
However, differences in relative potencies have
been demonstrated and a selective DA_2 antagonist,
domperidone, has been found.

An important practical reason for characterizing receptors
is to provide a basis for the development of new drugs. If
the molecular requirements for the activation of a receptor
and its specific antagonism can be elucidated, modifications
of agonist molecules can be made to produce compounds with qualitatively different pharmacological activity. In this paper
we describe two distinct dopamine (DA) receptors in peripheral
tissues, DA_1 receptors which subserve smooth muscle relaxation,
and DA_2 receptors which subserve inhibition of norepinephrine
release from the postganglionic sympathetic nerve. This subdivision is based on demonstration of pronounced differences in
the potency series of agonists active on both receptors, agonists
which act on only one of the two receptors, and selective antagonism (1).
    The DA_1 and DA_2 classification differs from other classifica-
tions based on anatomical locations (presynaptic/postsynaptic)
or response (increase in adenylate cyclase/no change in adenylate
cyclase; excitation/contraction). The subdivision of receptors
described in this paper also differs from subdivisions based
on the use of binding assays (2).

DA_1 receptors

Almost all investigations of the structure activity relationship (SAR) of DA_1 receptors were conducted in vascular preparations in which DA causes vasodilation in vivo and relaxation
in vitro. We have used essentially the same technique for investigating DA_1 agonists since 1965 (3,4). Renal blood flow is
measured by an electromagnetic flow meter in anesthetized dogs.
After administration of phenoxybenzamine, agonists are injected
directly into the renal artery and their effects on renal blood
flow are measured. Prior administration of an alpha-adrenergic
blocking agent is necessary since DA and most DA agonists cause
vasoconstriction by action on alpha-adrenergic receptors. Phenoxybenzamine does not significantly decrease the renal vasodilation produced by DA (5) as has been suggested by the inhibitory
action of phenoxybenzamine in a DA binding assay.
    Table I lists agonists which have been found to be active
in the renal vascular bed of the anesthetized dog. A few inactive
compounds are included for comparison. Each agonist was antagonized selectively by a DA antagonist in a dose which does not
affect vasodilation produced by bradykinin or isoproterenol.
Compounds shown in Table I were studied in our laboratory under
comparable conditions (2,4,5,7-10).

Table I

Structural requirements for DA_1 activity

Series	Substitution on nitrogen R_1	R_2	Approximate relative potency
3,4-Dihydroxy-phenylethylamines	H	H	1
	H	CH_3	1
	C_3H_7 (n)	C_3H_7 (n)	0.03
	C_3H_7 (n)	C_4H_9 (n)	0.04
	C_3H_7 (n)	C_5H_{11} (n)	0.04
	C_3H_7 (n)	phenethyl	0.007
	C_3H_7 (n)	C_2H_5	0.007
	C_4H_9 (n)	C_2H_5	0.04
2-Aminotetralins			
no hydroxy	C_3H_7 (n)	C_3H_7 (n)	inactive
5-hydroxy	C_3H_7 (n)	C_3H_7 (n)	inactive
7-hydroxy	C_3H_7 (n)	C_3H_7 (n)	inactive
5,6-dihydroxy	H	H	inactive
	C_3H_7 (n)	C_3H_7 (n)	1
6,7-dihydroxy	H	H	1
	H	CH_3	∿1
	C_3H_7 (n)	C_3H_7 (n)	0.25
	C_3H_7 (n)	C_4H_9 (n)	0.25

Continued next page

Table I continued

Structural requirements for DA_1 activity

Series	Substitution on nitrogen R_1	R_2	Approximate relative potency
1-Aminomethyl-5,6-dihydroxy isochromans	H	H	1
	H	CH_3	1
Aporphines	apomorphine		0.01 partial agonist
	isoapomorphine		inactive
	6-n-propyl norapomorphine		0.02
3',4'-dihydroxy-nomifensine			0.01 partial agonist
Benzazepines	SK&F 38393[a]		0.01 partial agonist
	SK&F 82526[b]		9

[a] 2,3,4,5-tetrahydro-7,8-dihydroxy-1-phenyl-1H-3-benzazepine

[b] 6-chloro-4'-hydroxy analog of SK&F 38393

The presence of two OH groups separated from the nitrogen atom by a distance of 7-8 angstrom units is a common feature of all active compounds. We had hypothesized previously that the beta rotameric form of DA, as typified by A-6,7-DTN, was the preferred conformation. However, we recently reported that dipropyl A-5,6-DTN was a potent DA_1 agonist (10) and, thus, the rotameric conformation may be of secondary importance. The R enantiomer of A-6,7-DTN is the active form (11). All of the agonists listed in Table I act on other receptors in addition to the DA_1 receptor. Important exceptions are the benzazepine derivatives, SK&F 38393 (12,13), and SK&F 82526 which act selectively on DA_1 receptors (14,15). The R form is the more active enantiomer (16).

DA_1 antagonists

DA_1 antagonists are studied in the same phenoxybenzamine-pretreated dog preparation as DA_1 agonists. We compare relative potencies of DA_1 antagonists by making simultaneous intra-arterial injections of DA with the antagonist under study. When administered intra-arterially in this way, most antagonists have a very short duration of action and many compounds can be studied in one experiment. The intra-arterial route of administration also permits estimation of responses prior to recirculation (2,4,17).
In addition, we calculate the relative specificity of the antagonists by determining the minimum dose required to attenuate bradykinin or isoproterenol divided by the minimum dose required to attenuate DA. As noted in Table II, sulpiride is the most potent antagonist and also has a large range of specificity. In contrast, haloperidol and the phenothiazines are much less potent and exhibit a very narrow range of specificity.
Because of the wide range of specificity of sulpiride and other substituted benzamides, it is possible to administer these drugs intravenously and differentiate renal or mesenteric vasodilation produced by DA from non-specific vasodilating agents such as bradykinin (18,19). However, we disagree with Shepperson et al (20) who estimated the relative potency of less specific antagonists by the intravenous route.

DA_2 agonists

Several techniques have been employed to assess the effects of DA_2 agonists on the postganglionic sympathetic nerve. For routine screening of DA_2 agonists we inject the compound into the femoral artery of the anesthetized dog and compare the resultant vasodilation with that produced by the standard agonist, dipropyl dopamine (DPDA) (21). Buylaert and his associates use apomorphine as the standard agonist. The vasodilation resulting from DPDA and apomorphine is known to be due to inhibition

Table II

DA$_1$ receptor antagonists with relative potencies and
ranges of specificity

Class	Antagonist	Relative potency (moles)	Range of specificity
Butyrophenones	Haloperidol	1.4×10^{-7}	<2
Phenothiazines	Chlorpromazine	2.5×10^{-7}	<2
	Prochlorperazine	2.5×10^{-7}	<2
	Trifluoperazine	2.5×10^{-7}	<2
	Fluphenazine	2.5×10^{-7}	<2
	Thioridazine	5.0×10^{-7}	<2
Substituted benzamides	Sulpiride	2.9×10^{-8}	>100
	Metoclopramide	1.5×10^{-6}	>10
	AHR 8812	2.8×10^{-8}	>50
	Bromopride	1.2×10^{-7}	>50
	Tiapride	2.8×10^{-7}	>50
	Sultopride	2.9×10^{-7}	>50
Others	Bulbocapnine	4.7×10^{-8}	∿8

*Minimal dose attenuating vasodilating responses to bradykinin
or isoproterenol divided by minimal dose attenuating vasodilating
responses to dopamine.

of sympathetic nervous system activity since it is eliminated by cutting the sympathetic nerves or by administering a ganglionic blocking agent (22). The vasodilation is specifically antagonized by DA_2 antagonists in doses which do not attenuate vasodilation produced by isoproterenol or bradykinin.

The advantage of utilizing femoral artery vasodilation in the intact dog as an endpoint is that the relative activity of compounds active on DA_1 and DA_2 receptors can be assessed under similar conditions in the anesthetized dog. The disadvantage of this technique is that compounds with potent alpha-adrenergic vasoconstrictor activity or non-specific vascular activity cannot be studied. However, compounds with alpha-adrenergic activity can be compared with standard agonists in the canine preparation of Long and associates (23). In this method the postganglionic sympathetic nerves are electrically stimulated and the reduction in tachycardia induced by electrical stimulation of the nerve is taken as a measure of DA_2 activity. Use of specific DA_2 antagonists is essential in this technique to separate the effects of DA_2 from alpha$_2$-adrenergic agonists.

In contrast to the relatively short list of compounds active as DA_1 agonists (Table I), a large number of compounds have been reported to be active on DA_2 receptors (2,11,24,25). Table III presents a selective list of these compounds. In addition to the demonstration that more compounds are active on the DA_2 receptors, pronounced differences in potency series have been demonstrated in agonists which are active on both receptors. For example, apomorphine is a weak partial agonist of DA_1 receptors, but is a potent full agonist of DA_2 receptors (2,22). Furthermore, DPDA, propylbutyl DA, propylpentyl DA, propylisobutyl DA, propylphenethyl DA, and propylethyl DA exhibit different potencies as DA_1 agonists, and all of them are much less potent than DA as DA_1 agonists. However, all these compounds are equipotent and strong agonists of the DA_2 receptor (21).

As with most DA_1 agonists, the majority of DA_2 agonists are active on other receptors. In particular, many ergot derivatives exhibit alpha-adrenergic and serotonin activity (26). In contrast, Bach et al (27) reported that the pyrazole derivative (LY 141865) is devoid of alpha and serotonin agonist and antagonist effects. Preliminary studies in our laboratories demonstrated that LY 141865 is a potent DA_2 agonist without alpha-adrenergic or DA_1 activity.

DA_2 antagonists

All compounds studied thus far as DA_1 antagonists also antagonize DA_2 receptors. We recently reported that domperidone is an extraordinarily selective DA_2 receptor antagonist (28). In a range of 0.5-5 µg/kg it produces dose related antagonism of DPDA induced femoral vasodilation without affecting the vasodilation produced by isoproterenol or bradykinin. In a dose of

Table III

Selected list of DA_2 agonists with relative DA_1 activity

Series	Compound	Relative activity	
		DA_2	DA_1
Dopamine (DA)	dimethyl DA	active	inactive
	dipropyl DA	all	weak
	propylbutyl DA	equipotent	and
	propylphenethyl DA	and	graded*
	propylethyl DA	active	
Aporphine	apomorphine	potent full agonist	weak partial agonist
6,7-Dihydroxy-2-aminotetralin (ADTN)	dimethyl ADTN	active	inactive
5,6-Dihydroxy-2-aminotetralin	primary amine	active	inactive
	N-N-dipropyl derivative	active	active
Ergot derivatives	bromocriptine	active	inactive
	pergolide	active	inactive
	lisuride	active	inactive
	lergotrile	active	inactive
Piribedil		active	inactive

*For relative activities, see Table I.

5 mg/kg intravenously, domperidone does not attenuate DA_1-induced renal vasodilation. Domperidone is ineffective as an antagonist of $alpha_2$-adrenergic receptors as determined by lack of antagonism of bradycardic effects of clonidine in the canine cardiac post-ganglionic nerve preparation.

Qualitative differences have also been observed in relative DA_1 and DA_2 activity of several antagonists. The differences are strikingly demonstrated with the enantiomers of sulpiride. (S)-sulpiride is much more active than (R)-sulpiride as a DA_2 antagonist; whereas, (R)-sulpiride is slightly more active than (S)-sulpiride as a DA_1 antagonist (11).

Spectrum of receptor activity of DA agonists

The determination of the relative activity of DA agonists on DA and other receptors is of critical importance in the elucidation of the structural requirements for synthesis of selective DA_1 and DA_2 agonists. For example, it should be possible to compare compounds with mixed receptor activity with the selective DA_1 agonist SK&F 82526 and the relatively selective DA_2 agonist, LY^141865 to elucidate the chemical requirements for the differential receptor activity (Figure 1).

Determination of relative receptor activity is also important for developing new DA agonists for cardiovascular and renal therapy. Fortunately, the results obtained in the anesthetized dog are relevant for human studies. Indeed, the clinical studies of DA (4) and propylbutyl DA (29) were direct extensions of canine investigations.

We compare the relative activity of DA_1 agonists on the following receptors: $beta_1$ (both directly and indirectly by release of norepinephrine from myocardial storage sites), $beta_2$, alpha, and DA_2. Table IV describes the spectrum of activity of several compounds studied in the pentobarbital anesthetized dog. The methods used have been described in detail (21,30).

Comparison of data obtained in vivo and in vitro

A comprehensive reivew of studies of the action of DA_1 agonists and antagonists in isolated blood vessels and organs was recently published (31). In general, there is good correlation between in vivo and in vitro data with isolated canine blood vessels. However, exceptions have been reported with isolated preparations of other species. The major differences are that sulpiride and its enantiomers appear to be much more potent in vivo than in vitro. Another discrepancy is that bromo-criptine, which is inactive on DA_1 receptors in vivo, has been shown to cause DA-like vasodilation of the isolated perfused rat kidney and relaxation of rabbit mesenteric artery strips. These data suggest the possibility that both DA_1 and DA_2 receptors may occur on vascular tissue in some species. However, other

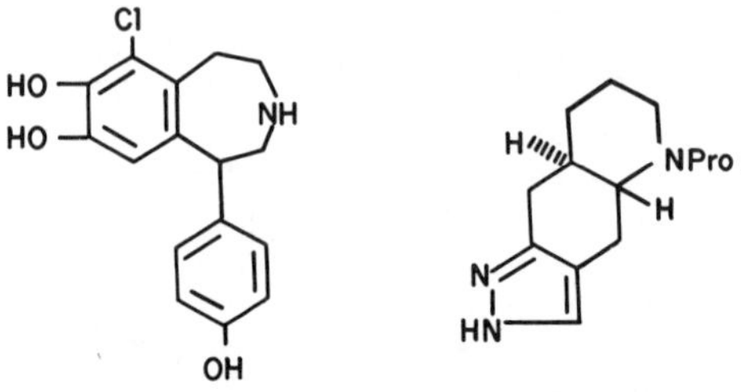

SK & F 82526   **LY 141865**

Figure 1.   The DA$_1$ agonist—SK&F 82526—and the DA$_2$ agonist—LY 141865.

Table IV

Spectrum of cardiovascular receptor activity of
selected DA agonists

	β$_1$/Indirect*		β$_2$	α	DA$_1$	DA$_2$
Dopamine (DA)	++	50%	0	++	+++	++
Epinine	+++	25%	+	+++	+++	++
A-6,7-DTN	++	100%	0	++++	+++	++
A-5,6-DTN	0	0	+++	0	0	++
Dipropyl-A-5,6-DTN	0	0	0	++++	+++	+++
Dipropyl DA	0	0	0	+	++	++
Propylphenethyl DA	0	0	0	+	+	++
SK&F 38393	0	0	0	0	++++	0
SK&F 82526	0	0	0	0	++++	0

*β$_1$ activity caused by release of norepinephrine from myocardial
storage sites.

possible mechanisms will have to be ruled out. The major problem
in comparing DA_2 receptors studied by the different techniques
is that $alpha_2$-adrenergic activity, which produces similar inhibition of norepinephrine release as activation of DA_2 receptors,
has not been clearly ruled out in many of the studies.

Determinations of the DA receptor subtypes responsible for physiological and pharmacological actions

In 1978 we compared the structure activity requirements
for the DA_1 (vascular DA) receptor with other putative DA receptors (1,4). The limited SAR data available at that time supported
the existence of at least two different DA receptors. DA-induced
responses were divided into two categories on the basis of differences in agonist activity and potency series. The following
phenomena appeared to be due to activation of DA_2 receptors:
inhibition of norepinephrine release from the postganglionic
sympathetic nerve; inhibition of ganglionic transmission; inhibition of prolactin release; and emesis. Review of the current
literature has not revealed exceptions to these correlations.
In contrast, only vasodilation has been clearly related
to DA_1 receptors. Interestingly, the potency series of agonists
for stimulation of adenylate cyclase and renal vasodilation
is similar (2). Furthermore, SK&F 82526 is active in both models
(15). However, no correlation was found in the relative actions
of antagonists (8). First, sulpiride, which is a potent antagonist of DA-induced renal vasodilation, is inactive as an antagonist of DA-induced stimulation of adenylate cyclase (32). Second,
ergot derivatives, which are inactive as DA_1 agonists, are antagonists of DA-induced stimulation of adenylate cyclase (33).
These exceptions prevent us from using the D_1 and D_2 subdivision
of Kebabian and Calne (34).
We have been unable to relate the DA_1 and DA_2 subtypes
to subdivisions proposed by radioligand binding assays. This
is an extremely controversial area: one to four different receptor subtypes have been postulated by binding assays. The problem
with binding assays appears to be related in part to lack of
selectivity of the ligands used. For example, spiperone has
been shown to bind to serotonin and non-specific binding sites
(36). In contrast, when the selective DA_2 antagonist, domperidone, is used as the ligand, a better relationship with DA_2
receptors can be demonstrated (36). A question to be answered
by future research is whether use of selective DA_1 ligands will
reveal a potency series of agonists similar to the DA_1 subtype.
Both DA_1 and DA_2 agonists cause behavioral changes and thus
demonstration of DA_1 receptors in central nervous system binding
assays should be possible if this method is valid (15,16,37).

Finally, behavioral models cannot be used to establish quantitative potency series of DA agonists and antagonists because of differences in blood-brain barrier penetration, regional distribution, and spectra of receptor and non-receptor actions (38). Hopefully, better correlation of peripheral and central receptors will occur as behavioral models improve and more selective agonists and antagonists become available.

Acknowledgments

We wish to gratefully acknowledge the cooperation of many chemists who provided us with compounds for study. We would also like to thank Ms Dana Glock for technical assistance and Ms Patricia Gomben for secretarial assistance. This work was suppoted by NIH grant GM-22220.

Literature Cited

1. Goldberg, L.I.; Kohli, J.D. Commun. Psychopharmacol. 1979, 3, 447-56.
2. Goldberg, L.I.; Volkman, P.H.; Kohli, J.D. Ann. Rev. Pharmacol. Toxicol. 1978, 18, 57-79.
3. McNay, J.L.; Goldberg, L.I. J. Pharmacol. Exp. Ther. 1966, 151, 23-31.
4. Goldberg, L.I. Pharmacol. Rev. 1972, 24, 1-29.
5. Goldberg, L.I.; Sonneville, P.F.; McNay, J.L. J. Pharmacol. Exp. Ther. 1968, 163, 188-97.
6. Creese, I.; Sibley, D.R. Biochem. Pharmacol. 1982, 31, 2568-9.
7. Dolak, T.M.; Goldberg, L.I. Ann. Reports Medicinal Chemistry 1981, Chpt. 11, 103-11.
8. Goldberg, L.I.; Kohli, J.D. "Advances in the Biosciences"; Pergamon Press, New York, 1982; p. 41-9.
9. Kohli, J.D.; Goldberg, L.I. J. Pharmacy Pharmacol. 1980, 32, 225-6.
10. Kohli, J.D.; Goldberg, L.I.; McDermed, J.D. Eur. J. Pharmacol. 1982, 81, 293-9.
11. Goldberg, L.I.; Kohli, J.D.; Listinsky, J.J.; McDermed, J.D. "Catecholamines: Basic and Clinical Frontiers"; Pergamon Press, New York, 1979; p. 447-9.
12. Pendleton, R.G.; Samler, L.; Kaiser, C.; Ridley, P.R. Eur. J. Pharmacol. 1978, 51, 19-28.
13. Hahn, R.A.; Wardell, J.R., Jr. J. Cardiovasc. Pharmacol. 1980, 2, 583-93.
14. Weinstock, J.; Wilson, J.W.; Ladd, D.L.; Brush, C.K.; Pfeiffer, F.R.; Kuo, G.Y.; Holden, K.G.; Yim, N.C.F.; Hahn, R.A.; Wardell, J.R., Jr.; Tobia, A.J.; Setler, P.E.; Sarau, H.M.; Ridley, P.T. J. Med. Chem. 1980, 23, 973-5.

15. Hahn, R.A.; Wardell, J.R., Jr.; Sarau, H.M.; Ridley, P.T. J. Pharmacol. Exper. Therap. 1982 (in press).
16. Kaiser, C.; Dandridge, P.A.; Garvey, E.; Hahn, R.A.; Sarau, H.M.; Setler, P.E.; Bass, L.S.; Clardy, J. J. Med. Chem. 1982, 25, 697-703.
17. Kohli, J.D.; Glock, D.; Goldberg, L.I. "The Benzamides: Pharmacology, Neurobiology, and Clinical Aspects"; Raven Press, New York, 1982; p. 97.
18. Glock, D.; Goldberg, L.I.; Kohli, J.D. Fed. Proc. 1981, 40, 290 #318.
19. Hahn, R.A.; Wardell, J.R., Jr. Arch. Pharmacol. 1980, 314, 177-182.
20. Shepperson, N.B.; Duval, N.; Massingham, R; Langer, S.Z. J. Pharmacol. Exp. Ther. 1982, 221, 753-61.
21. Kohli, J.D.; Weder, A.B.; Goldberg, L.I.; Ginos, J.Z. J. Pharmacol. Exp. Ther. 1980, 213, 370-4.
22. Buylaert, W.A.; Willems, J.L.; Bogaert, M.G. J. Pharm. Pharmacol. 1978, 30, 113-5.
23. Long, J.P.; Heintz, S.; Cannon, J.G.; Kim, J. J. Pharmacol. Exp. Ther. 1975, 192, 336-42.
24. Lokhandwala, M.F.; Tadepalli, A.S.; Jandhyala, B.S. J. Pharmacol. Exp. Ther. 1979, 211, 620-5.
25. Barrett, R.J.; Lokhandwala, M.F. Eur. J. Pharmacol. 1982, 77, 79-83.
26. Berde, B.; Schild, H.O. "Ergot Alkaloids and Related Compounds"; Springer-Verlag, New York, 1978.
27. Bach, N.J.; Kornfeld, E.C.; Jones, N.D.; Chaney, M.O.; Dorman, D.E.; Paschal, J.W.; Clemens, J.A.; Smalstig, E.B. J. Med. Chem. 1980, 23, 481-91.
28. Glock, D.; Kohli, J.D.; Goldberg, L.I. Fed. Proc. 1982, 41, 1651 #8077.
29. Fennell, W.; Taylor, A.; Brandon, T.; Goldberg, L.; Ginos, J.; Mitchell, J.; Miller, R. Clin. Res. 1980, 28, 469A.
30. Meyer, M.B.; Goldberg, L.I. Cardiologia 1966, 49, 1-10.
31. Brodde, O.E. Life Sci. 1982, 31, 289-306.
32. Spano, P.F.; Stefanini, E.; Trabucchi, M.; Fresia, P. "Sulpiride and other Benzamides"; Italian Brain Research Foundation Pres, Milan, Italy, 1979; pp. 11-31.
33. Kebabian, J.W.; Calne, D.B.; Kebabian, P.R. Commun. Psychopharmacol. 1977, 1, 311-8.
34. Kebabian, J.W.; Calne, D.B. Nature 1979, 277, 93-6.
35. Seeman, P. Biochem. Pharmacol. 1982, 31, 2563-8.
36. Lazareno, S.; Nahorski, S.R. Eur. J. Pharmacol. 1982, 81, 273-85.
37. Costall, B.; Naylor, R.J. Life Sci. 1981, 28, 215-29.

RECEIVED February 8, 1983

Commentary: Utility and Problems in the Classification of Dopamine Receptors

BARRY A. BERKOWITZ

Smith Kline & French Laboratories, Department of Pharmacology, Research & Development Division, Philadelphia, PA 19101

Dopamine and dopamine receptors have emerged as intriguing and useful target sites for chemical, biological, and therapeutic endeavors. This, when coupled with advances in physiological and receptor technologies, has led to a number of viewpoints on the classification of dopamine receptors. The brain and cardiovascular/renal systems have been the areas where the most work has been accomplished. Dopamine receptor studies in the central nervous system have been primarily biochemical, whereas those in the periphery have been primarily physiological. In their review Goldberg and Kohli have well summarized the present state of the art for physiologic and pharmacologic analysis of peripheral dopamine receptors and provide a reasonable classification designated DA$_1$ and DA$_2$ to identify the two major receptor subtypes. This commentary addresses the utility and problems of dopamine receptor classification.

The studies of Goldberg and colleagues have been and remain pioneering in not one but at least two fronts. First, the concepts and demonstration of vascular dopamine receptors has allowed and stimulated detailed studies of the location, function, mechanism, and role of peripheral dopamine and dopamine receptors. Second, their work led to and accelerated the utilization of dopamine agonists in clinical medicine with the use of intravenously administered dopamine for shock and heart failure being the best example.

In their communication the evidence is summarized suggesting two distinct types of dopamine receptors in the periphery, designated DA$_1$ and DA$_2$. Goldberg and Kohli's classification is based primarily on in vivo results using the vasculature and flow of blood to the renal, femoral and other selected

0097–6156/83/0224–0114$06.00/0

peripheral vascular beds. Most of the other classifications for dopamine receptors have been based on results obtained with the brain and cannot be assumed to apply to the vasculature and periphery. Interestingly, it is usually not the originators of any specific classification system who misuse or misapply it but frequently others who try to apply it to different areas and find exceptions.

Why characterize dopamine receptors?

There are a number of reasons to try to characterize dopamine receptor in addition to the fact that it is fashionable. Two of the major reasons well described by Goldberg and Kohli are (1) the importance of dopamine receptors subtypes responsible for physiological and pharmacological actions and (2) the utilization of the concept of selective dopamine receptors in the design and characterization of new drugs for cardiovascular and renal therapies.

Four Major Reasons for Problems in the Classification and Identification of Dopamine Receptors in the Periphery

Heterogeniety and structure of the vasculature. The multitude of functions of the vasculature, including conduit resistance and filtration vessels, may unfortunately be subserved by an equally complex heterogeniety of receptors and receptor locations. This obviously includes dopamine receptors. Unfortunately (or fortunately, depending on one's point of view) the vasculature is not "mushy" like the brain. The collagen and connective tissue content of blood vessels does not allow the gentle homogenization and tissue disruption which has allowed extensive biochemical analysis and binding studies of the brain dopamine receptor(s). Thus, there is a huge void in the biochemical analyses of the peripheral dopamine receptor(s) and mechanisms. There is clearly much to be done in this area for the future.

Dopamine has multiple actions on the cardiovascular system. Unfortunately dopamine is not a selective drug for dopamine receptors. Whereas we may strive to define actions of drugs or neurotransmitters as dopaminergic, these end points have been frequently defined under less than ideal conditions.

Dopamine also has prominent effects on α- and β-adrenoreceptors. Thus, in order to define any type of dopamine receptor or subtype, most in vivo or in vitro studies of the periphery and cardiovascular system must utilize a veritable pharmacopeia of drugs to block other receptors or mediator.

In vivo, studies with dopamine agonists must generally be performed using phenoxybenzamine or other α-adrenoreceptor blocking agents. In vitro, not only must α-blocking drugs be

used, but frequently, β-receptors blocker as well as cyclo-
oxygenase inhibitors. While this approach has been necessary,and is by no means unique for dopaminergic pharmacology, it
has yielded data which will require confirmation once we obtain
more selective drugs. The fact that most in vivo protocols
differ substantially in their use of drugs from in vitro protocols may well explain many of the conflicts in comparing dopaminergic actions in vivo and in vitro.

Lack of selective drugs active at dopamine receptors.
Obviously, there is a need for selective agonist and antagonists
of dopamine receptors. Almost by definition one cannot accu-
rately describe a receptor without a selective agonist or antagonist. This has been the difficult case we face with dopamine
receptor science.

Lack of In Vitro Model System. Whereas there are a variety
of in vitro receptor models for α- and β-adrenoreceptors there
has been less progress on well accepted models for dopamine
receptors. Tissue culture may well be used increasingly in the
future as a source of receptor material. In addition, there
needs to be continued work on the use of isolated vascular tissue to probe dopamine receptors.

Conclusion

    The tendency to examine one's data with a calculator in one
hand and a Greek dictionary in the other with the investigator
poised to name a new receptor need not be overly encouraged.
Where possible, receptor classification and concepts based upon
biochemical, physiological and pharmacological evidence would
seem to serve us best. Goldberg and Kohli have well summarized
the present state of the art for the physiologic and pharmacologic analysis of peripheral dopamine receptors. They are to
be commended for their care and contribution in defining dopamine receptors. Their work stands as a reasonable and secure
base for future studies.

RECEIVED January 7, 1983

5

Dopamine Receptors in the Neostriatum: Biochemical and Physiological Studies

J. C. STOOF

Free University, Department of Neurology, Medical Faculty,
Van der Boechorststraat 7, 1081 BT Amsterdam, The Netherlands

Dopamine induces biochemical and physiological
effects in the mammalian neostriatum. The occurrence of a D-1 dopamine receptor (in the classification scheme of Kebabian and Calne) accounts for
the ability of dopamine to enhance cyclic AMP formation. The occurrence of a D-2 dopamine receptor
accounts for the ability of dopamine to inhibit
cyclic AMP formation brought about by stimulation
of a D-1 dopamine receptor. Dopamine receptors
mediate the regulation of (1) the release or turnover of acetylcholine (postsynaptic dopamine receptor) and (2) the release or turnover of dopamine (presynaptic autoreceptor). Both receptors
can be classified as D-2 dopamine receptors. Indications for the occurrence of dopamine receptors
affecting the release or turnover of GABA, glutamate, serotonin and several neuropeptides are
evaluated.

Despite the recent burst of interest in the peripheral
actions of dopamine, this catecholamine is still best known as a
neurotransmitter in the central nervous system. This chapter
discusses the biochemical and physiological actions of dopamine
in the neostriatum and the substantia nigra, the regions containing most of the dopamine in the brain. The goal of this
chapter is to show that the concepts derived from the simple
peripheral systems containing a single category of dopamine receptor (1, 2, 3) can be applied to the more complex CNS. This
chapter is not intended to be an all inclusive compendium of
every technique used to study central dopamine receptors. Although dopamine receptors can be studied in binding assays or by
behavioral protocols, I will not focus attention on either of
these methodologies. Binding studies of dopamine receptors, although easy to perform, have yielded too many data, too many

0097–6156/83/0224–0117$08.50/0

categories of dopamine receptors and too many controversies (4, see also Caron, this volume). In part, this situation may be a consequence of the sensitivity of binding assays to minor changes in assay conditions (5). The interpretation of behavioral studies is hampered by the current ignorance of the neuroanatomy and neurophysiology as to how stimulation of a dopamine receptor is translated into an observable behavior. This chapter will focus attention upon the dopamine receptors in the nigro-neostriatal axis which elicit either biochemical or physiological events amenable to in vitro experimental investigation. In discussing the pharmacology of each system, I will use the two dopamine receptor hypothesis as the basis for my consideration of receptor pharmacology (see Brown and Dawson-Hughes & Kebabian et al., this volume).

The Nigro-Neostriatal Dopaminergic Neurons

Some 3500 dopaminergic neurons located in the zona compacta of the substantia nigra innervate the entire neostriatum (6, 7). Within the neostriatum, the axons of these neurons form an extremely dense terminal arborization that can be visualized with fluorescence histochemistry and immunocytochemical techniques (8, 9). This terminal arborization contains approximately 1 billion dopaminergic varicosities and forms about 20% of all the varicosities (10) present in the neostriatum.

Postsynaptic Dopamine Receptors

The majority of neurons in the neostriatum are interneurons. The neostriatum receives afferents from diverse areas like the thalamus, the dorsal raphe nuclei, the cerebral cortex and the substantia nigra. Efferents from the neostriatum innervate the globus pallidus and the substantia nigra (11). The neostriatum contains many neurotransmitters and putative neurotransmitters including: acetylcholine, dopamine, γ-aminobutyric acid, glutamate, serotonin and the peptides cholecystokinin, enkephalin and substance P (12, 13). Theoretically, the dopaminergic neurons could communicate via dopaminergic synapses and postsynaptic dopamine receptors with neurons containing each of these neurotransmitters. Consequently, changes in either the release or the turnover of any of these neurotransmitters could provide evidence of activity at a postsynaptic dopamine receptor.

A Postsynaptic Dopamine Receptor Regulating the Release or Turnover of Acetylcholine

Acetylcholine is a major neurotransmitter in the neostriatum. Within the neostriatum, both the content of acetylcholine and the specific activity of choline acetyltransferase,

the enzyme synthesizing acetylcholine, are extremely high (14).
Most of the neostriatal acetylcholine occurs within interneurons; only a small amount of neostriatal acetylcholine can be
associated with afferent neurons originating in the center
median-parafascicular complex of the thalamus (11).

The dopaminergic nigro-neostriatal neurons make synaptic
contacts with the cholinergic interneurons (15, Figure 1). The
content of acetylcholine in the neostriatum is increased by
dopamine receptor agonists and decreased by dopamine receptor
antagonists (16, 17, 18). Apomorphine and L-DOPA inhibit the
in vivo release of acetylcholine from cat neostriatum; this
effect is blocked by neuroleptics (19). RU 24926 (chemical
structure depicted in Figure 2), a selective D-2 agonist, increases the content of neostriatal acetylcholine (20). Together,
these data from in vivo experiments are consistent with the
hypothesis that stimulation of a D-2 receptor upon the cholinergic interneurons increases the content of neostriatal acetylcholine by blocking its release.

In vivo experiments studying either the content or the
release of neostriatal acetylcholine are time consuming, technically difficult and not always easy to interpret. Conversely,
in vitro studies of acetylcholine release with a superfusion
technique are less time consuming, technically easier to perform
and easier to interpret. With this in vitro technique, slices of
neostriatum are incubated with radiolabeled choline, a precursor
of acetylcholine, in order to label the pool of newly synthesized acetylcholine. The tissue is transferred to a superfusion
apparatus and the calcium-dependent release of radiolabeled
acetylcholine is evoked electrically, with veratridine or with
elevated potassium concentrations (21). The superfusion technique is a simple procedure for investigating the release of
acetylcholine (or other neurotransmitters). However, there are
some limitations to the technique. For example: dopaminergic
agonists elicit no more than a 50% to 60% inhibition of acetylcholine release from rat neostriatum (e.g. Figure 3). It is not
clear if this implies that 40% of the cholinergic interneurons
in rat neostriatum do not possess dopamine receptors. Alternatively, because different species (rat versus cat or rabbit)
and different techniques (potassium-evoked release versus electrically stimulated release) give quantitatively different
results (22-27), it remains possible that there are technical
limitations to the precision of this technique. A more detailed
description of this method is presented elsewhere (26, 27).

The results obtained from in vitro superfusion experiments
are in accord with the conclusion that the cholinergic interneurons possess a D-2 dopamine receptor. Dopamine inhibits the
release of [3H]-acetylcholine from neostriatal tissue; however,
concentrations greater than 1 μM are required to achieve maximal
inhibition. Because dopamine is removed from the extracellular
space by the dopaminergic nerve terminals, it is difficult to

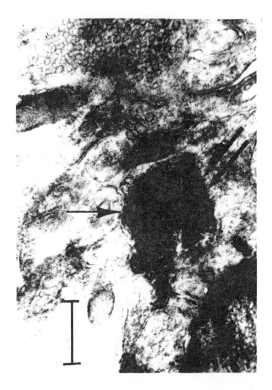

Figure 1.   Degenerating nerve ending (→) in guinea pig neostriatum following 6-OH-dopamine administration making asymmetrical synaptic contact with dendritic spine positively staining for cholineacetyltransferase (⇒). Bar indicates 0.125 μm. (Reproduced with permission from Ref. 15. Copyright 1976, Elsevier Biomedical Press.)

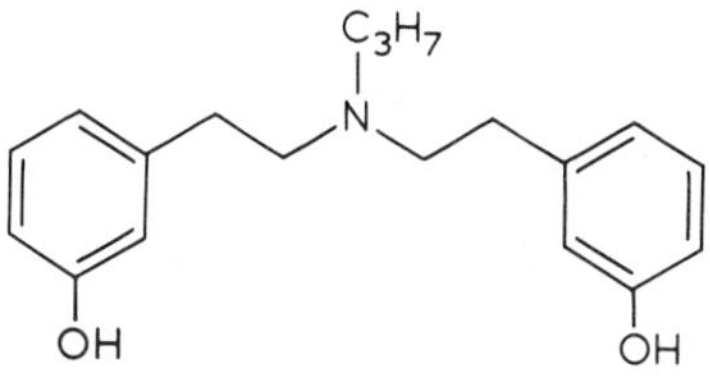

RU 24926

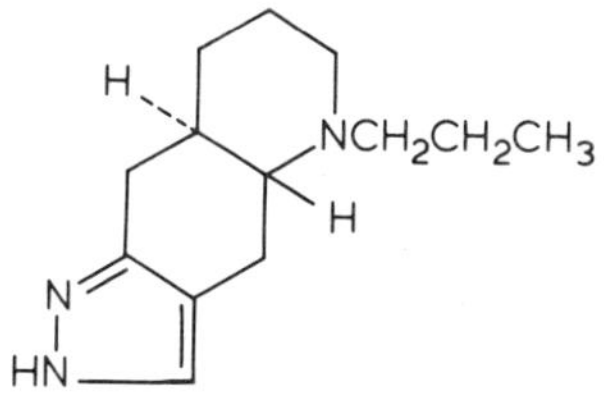

LY 141865

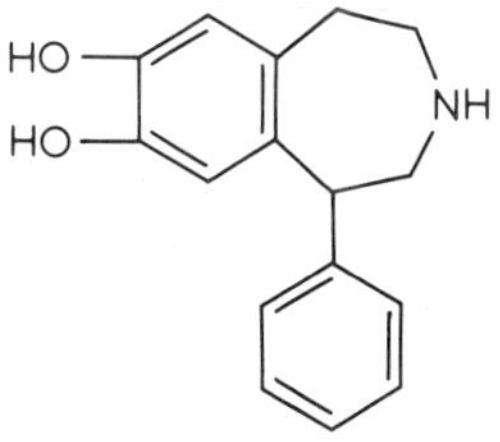

SK&F 38393

Figure 2.   Chemical structures of LY 141865, RU 24926, and SKF 38393.

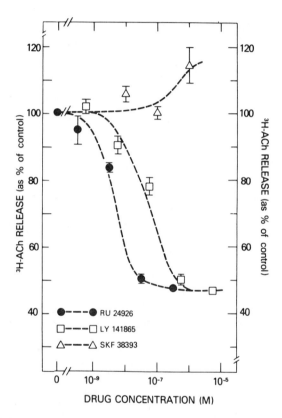

Figure 3. *D-2 receptor agonists, RU 24926 and LY 141865, inhibit the K$^+$-stimulated release of [^{3}H]-acetylcholine from blocks of rat neostriatum. (Reproduced with permission from Ref. 96. Copyright, Elsevier Biomedical Press.)*

reliably estimate the concentration of dopamine causing maximal
inhibition. Apomorphine and aminotetralin derivatives inhibit
in a dose-dependent manner the release of [3H]-acetylcholine
from neostriatal tissue (22-27). The selective D-2 agonists,
LY 141865 and RU 24926 (28, 29; chemical structures depicted in
Figure 2) inhibit the release of [3H]-acetylcholine from neostriatal tissue slices (Figure 3). The maximal effect of either
drug is an approximate 50% reduction of the fractional rate of
release of [3H]-acetylcholine. LY 141865 is half-maximally
active at a concentration of 70 nM; this is very similar to its
potency upon the D-2 receptor in the intermediate lobe of the
rat pituitary gland (28). RU 24926 is half-maximally active at
a concentration of 7 nM; this is very similar to its potency
upon the D-2 dopamine receptor on the mammotrophs of the anterior pituitary gland (29). Likewise (-)-sulpiride, a selective
D-2 antagonist (3, 30, 31), reverses the inhibition of
[3H]-acetylcholine release induced by either LY 141865 or
RU 24926 (Figure 4). Furthermore, SKF 38393, a selective D-1
agonist (32, 33, 34; chemical structure depicted in Figure 2),
does not inhibit (and at high concentrations slightly stimulates) the release of [3H]-acetylcholine (Figure 3). Other
neuroleptic drugs antagonize the inhibitory effect of dopamine
and dopaminergic agonists on the release of [3H]-acetylcholine
(22, 26, 27). The in vitro effects of dopaminergic agonists and
antagonists upon the release of [3H]-acetylcholine reinforce the
conclusion drawn from in vivo studies that a D-2 receptor regulates the release of neostriatal acetylcholine.
    Despite the technical or methodological limitations of
experiments determining the release of acetylcholine from the
neostriatum, this experimental parameter is an extremely valuable model for studying a postsynaptic CNS dopamine receptor.
Acetylcholine release is one of the few physiological parameters
regulated by a dopamine receptor that can be quantified with in
vitro techniques. Furthermore, dopaminergic regulation of
acetylcholine release is a matter of some practical interest.
For example, Parkinson's disease is a neostriatal dopamine deficiency syndrome (35). The loss of neostriatal dopamine disrupts
the balance between the neostriatal dopaminergic and cholinergic
systems that is thought to regulate normal activity in the neostriatum (36). The dopaminergic agonists presently used in the
treatment of Parkinsonism (37, 38) may achieve some of their
therapeutic effect by directly stimulating dopamine receptors;
however some of their therapeutic effect may be a consequence of
stimulating the dopamine receptor upon the cholinergic inter-
neuron and thereby restoring the balance between the dopaminergic and cholinergic systems in the neostriatum. Prior to the
advent of L-DOPA therapy for Parkinsonism, cholinergic antagonists were widely used to alleviate the symptoms of this
disease (39).

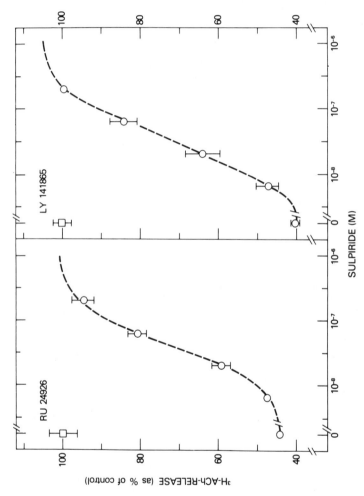

Figure 4. *Reversal by (−)-sulpiride of either the RU 24926 (left) 0.1 μM or the LY 141865 (right) (1.0 μM)-inhibited release of [³H]-acetylcholine (K⁺-stimulated) from blocks of rat neostriatum. The release of [³H]-acetylcholine was determined without drugs (☐) or with one of the D-2 receptor agonists in the presence of (−)-sulpiride (○). (Reproduced with permission from Ref. 96. Copyright, Elsevier Biomedical Press.)*

A Postsynaptic Dopamine Receptor Regulating the Release
or Turnover of GABA?

Both GABA and glutamic acid decarboxylase (GAD), a marker
enzyme for GABA-containing nerve terminals, occur in the neostriatum, globus pallidus and substantia nigra (12). The somata
of GABA-containing neurons occur predominantly in the globus
pallidus: these neurons project to the zona reticulata of the
substantia nigra; their recurrent collaterals project to the
neostriatum.GABA-containing interneurons and glia also occur in
the neostriatum (40-43).

  In vivo studies indicate that dopamine may influence the
GABA-containing cells in the neostriatum and substantia nigra.
Thus, apomorphine increases the content of neostriatal GABA
(44); conversely, haloperidol decreases the content (45) and
turnover (46) of neostriatal GABA. Chronic (8 weeks) treatment
with either haloperidol or chlorpromazine increases neostriatal
GAD activity but does not alter the content of GABA within the
neostriatum. Destruction of the nigro-neostriatal dopaminergic
neurons also increases neostriatal GAD activity (47). In experiments using a push-pull canula, either apomorphine or dopamine
inhibits the release of endogenous neostriatal GABA (48). In the
substantia nigra, haloperidol decreases the content of GABA
(45), while either apomorphine or dopamine inhibit the release
of endogenous GABA (49).

  The effects of dopaminergic drugs upon the synthesis,
storage and release of GABA are in accord with the possibility
that a dopamine receptor might regulate the activity of GABA-
containing neurons. However, this possibility can be accepted
with only limited enthusiasm. Extremely high doses of dopaminergic drugs are required to elicit effects in the neostriatal
GABA system. Although apomorphine at a dose of 0.05 mg/kg will
stimulate dopamine receptors (50), apomorphine must be used at a
dose of 20 mg/kg to induce a 25% increase in neostriatal GABA
(44). Likewise, haloperidol must be used at 10 mg/kg to induce a
37% decrease in the content of neostriatal GABA (45). Such a
massive dosis of haloperidol results in brain concentrations so
high that haloperidol can block not only dopamine receptors but
also α-1 and α-2 adrenoceptors, H-1 and H-2 histamine receptors
and serotonin receptors (51). Therefore, it appears optimistic
to conclude from such in vivo experiments that the observed
effects of dopaminergic drugs upon the neostriatal GABA system
result from an interaction with a specific dopamine receptor.
This negative conclusion is in accord with the data of Pycock et
al. (52) who could not demonstrate that subtle manipulations of
central dopaminergic systems altered the concentration of GABA.

  The possibility that dopamine regulates neostriatal GABA-
ergic function receives no support from in vitro superfusion
experiments determining the release of [3H]-GABA from slices of
rat neostriatum (53). The D-2 receptor agonist RU 24926 is

ineffective in modulating the potassium-stimulated release of
[3H]-GABA at concentrations maximally inhibiting acetylcholine
release (Figure 3). Also the D-1 receptor agonist SKF 38393 does
not cause a significant change in the release of [3H]-GABA from
neostriatal tissue. Therefore, the reported effects of 100 μM
apomorphine (48, 49, 54) must be explained as a consequence of
an interaction with an entity other than a D-1 or a D-2 dopamine
receptor.

As noted earlier, GABA appears to be a neurotransmitter in
the substantia nigra. Following chronic blockade of dopamine
receptors with haloperidol, the turnover rate of nigral GABA is
depressed; acute haloperidol is without effect (47). The effects
of dopamine upon the release of [3H]-GABA from the substantia
nigra are controversial. Reubi et al. (55) report that dopamine,
at high concentrations, stimulates the release of [3H]-GABA from
the nigra in vitro. Dibutyryl cAMP mimicks this effect of dopamine, thereby raising the possibility that a D-1 receptor mediates this effect. However, Arbilla et al. (56) could not re-
produce these observations and concluded that dopamine does not
modulate the spontaneous or stimulus-evoked release of GABA in
the substantia nigra. Therefore, it seems premature to accept
the possibility that dopamine affects GABAergic function in the
substantia nigra via receptors located on GABAergic neurons.

A Postsynaptic Dopamine Receptor Regulating the Release or Turnover of Glutamate?

The neostriatum receives a massive neuronal input from the
ipsilateral cerebral cortex (57). These fibers distribute, in an
organized way, to all parts of the neostriatum. The heaviest
projection comes from the sensorimotor cortex. Spencer (58) initially suggested that (part of) these corticostriatal fibers
utilize glutamate as a neurotransmitter. Evidence implicating
glutamate as the major neurotransmitter used by these fibers is
gradually accumulating (43). Cortical ablation has been found to
result in 40-50% reduction of high affinity uptake of labeled
glutamate in neostriatal tissue (59, 60). Striatal neurons are
excited by either direct cortical stimulation or by ionto-
phoretic application of glutamate; these effects are blocked by
L-glutamate diethylester, a glutamate antagonist (58). Electrical stimulation of the frontal cortex induces specific release
of labeled glutamic acid from rat neostriatum (61). Furthermore,
endogenous glutamate is released, in a calcium-dependent manner,
by elevated concentrations of potassium ions (62).

Dopamine receptors may occur on the terminals of the
glutamatergic afferents to the neostriatum (63). Following
cortical ablation, the number of [3H]-haloperidol binding sites
in the striatum is reduced by 32% (63). In an in vitro superfusion system, dopamine, apomorphine, amino tetralin derivatives
or bromocriptine inhibit the depolarization-induced release of

[3H]-glutamate (64, 65). However, very high concentrations of
these drugs are needed to elicit this inhibitory effect. In one
of these studies dibutyryl cAMP did not mimic the effects of the
dopaminergic agonists (65); therefore, it is very unlikely that
a D-1 receptor mediates this modulation of glutamate release. In
a recent study (53), neither the D-1 agonist SKF 38393, nor the
D-2 agonist RU 24926, altered the depolarization-induced release
of [3H]-glutamate (even when tested at concentrations maximally
active on their respective receptors). Thus, it appears that
neither a D-1 nor a D-2 receptor modulates the release of
glutamate.

A Postsynaptic Dopamine Receptor Regulating the Release or
Turnover of Serotonin?

Serotonin-containing neurons project from the raphe nuclei
to the neostriatum (66). Lesioning techniques show that this
projection originates only in the dorsal raphe nuclei (67, 68).
In addition, a major serotoninergic pathway from the medial and
dorsal raphe nuclei (69) projects to the substantia nigra. In
preparing this review, I did not encounter any reports suggesting that the turnover or release of serotonin in the neostriatum
is directly influenced by drugs stimulating or blocking dopamine
receptors. Furthermore, according to Hassler (43) and Pasik
(70), there are not many axo-axonal contacts in the neostriatum.
This makes the occurrence of direct synaptic contacts between
dopaminergic and serotoninergic nerve terminals unlikely.
Obviously, it is still possible that both neuronal systems
communicate via mechanisms other than synaptic contacts.
A few studies describe interactions between the dopaminergic and the serotoninergic systems in the substantia nigra,
thereby suggesting a role for nigral dopamine receptors.
Dopamine (0.1 μM) inhibits the release of [3H]-serotonin (71),
while apomorphine (50 μM) stimulates the release of [3H]-sero-
tonin (72). These apparently conflicting observations preclude
any definitive conclusions being drawn about the involvement of
dopamine receptors in regulating the turnover or release of
serotonin in the substantia nigra.

A Postsynaptic Dopamine Receptor Regulating the Release or
Turnover of Peptide Neurotransmitters?

The neostriatum contains many of the peptides which are
currently fashionable research entities (for a review see 73).
In this section, I will mention the peptides which have been
implicated as being under dopaminergic control. Substance P was
among the first peptides discovered in the nigro-neostriatal
axis. This peptide occurs in both interneurons and in neurons
projecting to the substantia nigra (74). Several pieces of
circumstantial evidence point to a dopaminergic effect upon

substance P. Chronic treatment with haloperidol reduces the
concentration of substance P in the substantia nigra (75).
However, treatment with apomorphine also lowers nigral substance
P concentration; haloperidol blocks this effect (76). Likewise,
amphetamine causes a dose-dependent reduction in substance
P-like immunoreactive material within the neostriatum (but not
in the substantia nigra); haloperidol blocks this effect (77).

The neostriatum has a high content of enkephalin and these
enkephalin-like peptides occur in neostriatal interneurons (78).
Chronic treatment with haloperidol increases both the content
and the release of neostriatal enkephalin (79). However, in
vitro studies fail to demonstrate any dopaminergic regulation of
enkephalin release (79).

Cholecystokinin (CCK)-like immunoreactive material occurs
in the CNS and the caudate has the highest concentration of any
brain region (13, 80). However, dopaminergic regulation of neostriatal CCK release has not been investigated (81). Interestingly, CCK-like peptide(s) coexist with dopamine in certain
dopaminergic neurons (73). This raises the possibility that
CCK might regulate dopaminergic activity with a novel, but at
present unknown, mechanism.

In summary, it is difficult to generalize about dopaminergic control of peptidergic function in the neostriatum. This
circumstance is a consequence of the ignorance of the physiological functions regulated by the peptides and the limited
number of investigations directed towards neostriatal peptides.

A Postsynaptic Dopamine Receptor Regulating Cyclic AMP Formation

A postsynaptic dopamine receptor in the neostriatum can be
characterized with biochemical procedures. The basis for this
characterization is the ability of dopamine to stimulate
adenylate cyclase activity in cell-free homogenates of neostriatal tissue. Earlier in this volume, Brown and Dawson-Hughes
discuss the properties of this receptor-enzyme system in the
bovine parathyroid gland and summarize the evidence that a
dopamine-stimulated formation of cyclic AMP triggers the release
of parathyroid hormone from this bovine tissue. However, the
role of this enzyme in either the neostriatum or the substantia
nigra is unknown. In the substantia nigra, dopamine was reported
to stimulate the release of [3H]-GABA, and this effect was
mimicked by dibutyryl cyclic AMP (55). However, as noted earlier
in this chapter, these observations could not be reproduced by
another group (56). Thus, in both the neostriatum and the substantia nigra, the dopamine-sensitive adenylate cyclase is a
receptor in search of a function.

At least a major part of the neostriatal dopamine-sensitive
adenylate cyclase activity is not associated with the terminals
of the dopamine-containing nigro-striatal neurons. Intranigral
injections of 6-OH-dopamine which destroy these dopamine-

containing neurons do not cause a loss of striatal dopamine-
sensitive adenylate cyclase activity (82). However, intra-
striatal injections of kainic acid which destroy neuronal somata
cause a substantial loss of this dopamine-sensitive enzyme
activity (83, 84). In studies of the subcellular distribution
of this enzyme activity, the highest specific enzyme activity
is found in submitochondrial fractions enriched with nerve
endings (85). These observations are compatable with a postsynaptic location of the enzyme activity upon interneurons or
recurrent collaterals of neostriatal efferents. A dopamine-
senstive adenylate cyclase activity also occurs in the substantia nigra (86, 87, 88). This nigral enzyme is not associated
with the dopamine-containing neurons.

The pharmacological properties of the dopamine-sensitive
adenylate cyclase activity in either the bovine parathyroid
gland or the neostriatum are summarized by Brown and Dawson-
Hughes earlier in this volume and by other authors elsewhere
(89, 90).

Dopamine, in addition to stimulating the formation of
cyclic AMP, is also able to inhibit the formation of this cyclic
nucleotide. This inhibitory effect of dopamine can be clearly
demonstrated in either the mammotrophs of the anterior pituitary
gland (91-94) or the melanotrophs of the intermediate lobe of
the pituitary gland (see Kebabian et al., this volume). Both
the stimulatory and the inhibitory effect of dopamine under
cyclic AMP formation can be demonstrated in the neostriatum
using in vitro superfusion. Either dopamine or SKF 38393
stimulates the efflux of cyclic AMP from slices of rat neostria-
tum(Figures 5 and 6,95, 96); this is in accord with the ability
of either of these drugs to stimulate adenylate cyclase activity
in cell-free homogenates of the neostriatum (37). LY 141865, the
selective agonist upon the D-2 dopamine receptor (28), reduced
the magnitude of the SKF 38393-induced efflux of cyclic AMP but
did not change the concentration of agonist giving half-maximal
efflux. The inhibitory effect of LY 141865 occurs even in the
absence of calcium ions from the superfusion medium. Since it
is commonly accepted that calcium ions are essential for the
release of neurotransmitter (21) this latter observation
suggests that neurotransmitter release is not required for the
dopaminergic inhibition of cyclic AMP release (96). In addition,
(-)-sulpiride, an antagonist of the D-2 dopamine receptor,
markedly potentiates the dopamine-stimulated efflux of cyclic
AMP but does not appreciably change the molar potency of
dopamine (Figure 6). It must be stressed here that in the
experiment depicted in Figure 6 an unusually high concentration
of (-)-sulpiride has been used (50 μM). However, one has to
realize that approximately 100 μM dopamine is required to
maximally activate the D-1 dopamine receptor. To block the
effects of 100 μM dopamine on the D-2 dopamine receptor high
concentrations of (-)-sulpiride are needed. Figure 7 presents a

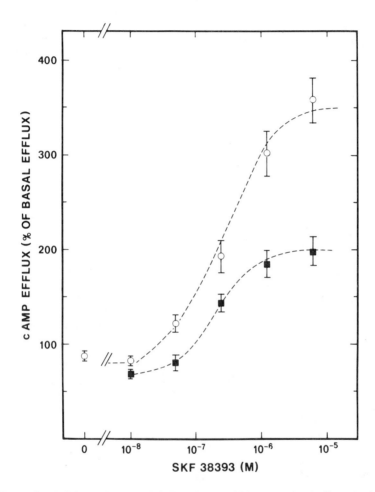

Figure 5.   Inhibition by LY 141865 of the SKF 38393-stimulated efflux of cAMP from blocks of rat neostriatum. The efflux of cAMP from neostriatal tissue, stimulated with the indicated concentrations of SKF 38393, was estimated in the absence (○) or presence (■) of 5 μM LY 141865. (Reproduced with permission from Ref. 96. Copyright, Elsevier Biomedical Press.)

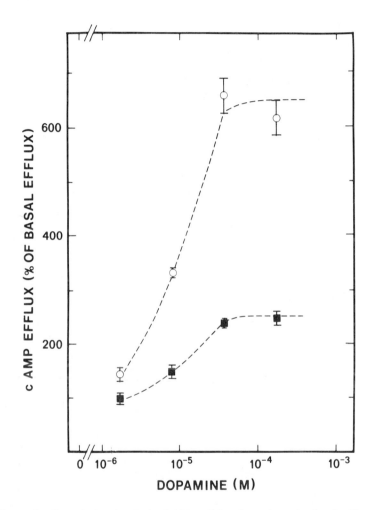

Figure 6.   Potentiation by (−)-sulpiride of the dopamine stimulated efflux of cAMP from blocks of rat neostriatum. The efflux of cAMP from neostriatal tissue, stimulated with the indicated concentrations of dopamine, was estimated in the absence (■) and presence (○) of 50 μM (−)-sulpiride. (Reproduced with permission from Ref. 96. Copyright, Elsevier Biomedical Press.)

diagramatic summary of a working hypothesis concerning the arrangement of the D-1 and D-2 receptors regulating cyclic AMP formation in the neostriatum.

Presynaptic Dopamine Receptors

The dopamine-containing nigro-neostriatal dopaminergic neurons possess receptors for dopamine. These dopamine receptors occur on both the nerve terminals within the neostriatum (presynaptic autoreceptors) as well as on the soma and dendrites (soma-dendritic autoreceptors). Stimulation of either category of autoreceptor can regulate the synthesis, turnover and release of dopamine in the neostriatum but by different mechanisms.

Autoreceptors Regulating the Turnover of Dopamine

Stimulation of the presynaptic autoreceptors inhibits tyrosine hydroxylase activity and dopamine synthesis within the dopaminergic nerve terminals in the neostriatum (97-101; however see 102 for a negative report). The effects of dopaminergic drugs which can be ascribed to an action upon the presynaptic autoreceptor include:
(1) Blockade of dopamine receptors by administration of neuroleptics has been demonstrated to induce an allosteric activation of tyrosine hydroxylase which is accompanied by an increase in the turnover rate of dopamine and accumulation of dihydroxyphenylacetic acid (DOPAC) and homovanillic acid (HVA) (103-106).
(2) Dopamine agonists produce a decrease of brain DOPAC and HVA concentrations, which can be antagonized by neuroleptic drugs (107, 108).
(3) Cessation of impulse flow in dopaminergic neurons can be achieved by administration of gammabutyrolactone (GBL). In the presence of a dopa decarboxylase inhibitor this induces an accumulation of DOPA. Administration of dopamine receptor agonists decreases this accumulation of DOPA, an affect which can be antagonized by pretreatment with neuroleptics (recently reviewed by Roth, 109).
Systemical or intranigral application of dopamine receptor agonists depresses the firing of nigrostriatal dopamine cells by stimulating soma-dendritic autoreceptors in the zona compacta, an effect again antagonized by neuroleptics (110, 111, 112).
Many of these effects can still be observed after pretreatment with kainic acid in the neostriatum (113). This treatment degenerates the neostriatal efferent pathways and by consequence also the neostriatal efferent part(s) of the nigrostriatal loop. Apparently postsynaptic dopamine receptors, if located on these striatonigral neurons play only a minor role in the regulation of dopamine synthesis and turnover in dopaminergic neurons and the major role has to be attributed to dopaminergic autoreceptors.

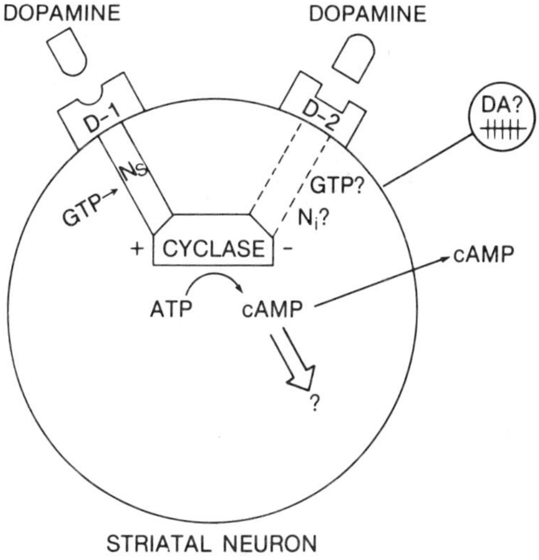

Figure 7. Hypothetical model for the dual regulation of adenylate cyclase activity by dopamine in a neuron of the rat neostriatum.

Presynaptic Autoreceptor Regulating the Release of Dopamine

The in vitro release of dopamine from brain structures receiving a dopaminergic input has been extensively studied. The release of dopamine from neostriatal tissue can be modulated by many substances including dopamine and dopaminergic drugs (for reviews see 114 and 115). Initially, data reported on the effects of dopaminergic drugs on dopamine release appeared inconsistent. A number of groups reported that dopamine receptor agonists inhibited the release of radiolabelled dopamine from striatal slices (116, 117, 118), while others did not observe such an inhibition of dopamine release (119, 120, 121). However, the possibility of demonstrating autoreceptor-mediated modula-. tion of transmitter release proofed to be critically dependent upon the experimental procedures used, particularly with respect to the conditions for stimulation of release. Dopamine-receptor agonists induced a much stronger inhibition of dopamine release at low frequency than at high frequency electrical stimulation (122). Similarly dopamine release induced by 20 mM K^+ could be inhibited by dopamine receptor agonists to a much greater extent than dopamine release induced by 40 mM K^+ (27).

At present it seems to be commonly accepted that under carefully selected conditions the release of dopamine in vitro can be inhibited by dopamine and other dopaminergic agonists (27, 116, 117, 118, 122-126). This inhibition can be antagonized by neuroleptic drugs. Although no indusputable evidence, these data are further indications for the presence of presynaptic autoreceptors on the membranes of the varicosities of dopaminergic neurons. Over a period of several years notably the groups of Langer and Starke in a series of studies have been able to steadily improve the conditions to see the presynaptic dopamine receptors "at work". They arrived at experimental conditions under which the release of dopamine was inhibited by about 90% by dopamine receptor agonists. Whether these conditions reflect physiological conditions is still an open question. Most of their studies have been carried out with neostriatal tissue from rabbits or cats and dopamine release was evoked electrically. Using neostriatal tissue from rats and 20 mM K^+ to evoke release of dopamine, we never found a higher inhibition than 40-45% induced by dopamine receptor agonists (27, 53). This difference could result from species differences or from differences in stimulating conditions. In a recent study Lehmann and Langer (122) demonstrated very clearly (Figure 8) that the extent of inhibition of dopamine release by the dopamine receptor agonist pergolide was dependent on the frequency of the electrical stimulation used. With increasing frequencies of stimulation, the maximal inhibition caused by pergolide decreased as anticipated. Surprisingly the IC_{50} of pergolide increased with increasing frequencies of stimulation: at a stimulation frequency of 6 Hz the IC_{50} for pergolide was roughly

20-fold the IC_{50} value obtained at 1 Hz. Thus it must be emphasized that the IC_{50} value of a dopamine receptor agonist may critically depend on the experimental conditions.

Pharmacological Characterization of Dopaminergic Autoreceptors

The autoreceptors mediating inhibition of the turnover of dopamine ressemble the D-2 receptor rather than the D-1 receptor. For example, LY 141865, the selective D-2 agonist, displays agonist activity upon the autoreceptor in either the GBL model or the synaptosomal tyrosine hydroxylase assay (127). In contrast, SKF 38393 has not been reported to be active in any experimental model of these autoreceptors. In several experimental models of the presynaptic autoreceptor, (-)-sulpiride displays antagonist activity (128-130). However, ergots which are potent D-2 agonists in the periphery are not agonists in the synaptosomal tyrosine hydroxylase assay although they are active in in vivo models of the autoreceptor (131). In preparing this review, I encountered no suggestion for a difference in the pharmacological properties of the presynaptic and the soma-dendritic autoreceptors; indeed, Roth argues that the two receptors are pharmacologically similar (109).

The presynaptic autoreceptor modulating dopaminergic release appears to be a D-2 receptor as well. LY 141865 inhibits the potassium-evoked release of dopamine (Figure 9); conversely SKF 38393 is without effect upon dopamine release. Other dopaminergic agonists inhibiting dopamine release include apomorphine, ergots and amino tetralin derivatives. Sulpiride and other benzamides as well as butyrophenones and phenothiazines antagonize the inhibitory effect of dopamine and other dopaminergic agonists (27, 114-118, 122-126, 132).

A Presynaptic Autoreceptor Regulating cyclic AMP Formation?

Activation of neostriatal tyrosine hydroxylase was observed when cyclic AMP was added to high speed supernatants from rat neostriatum (133). Intraventricular injection of dibutyryl cyclic AMP stimulated tyrosine hydroxylation in the neostriatum (134). However, it is still questionable if under physiological conditions this cyclic AMP involvement in the feedback control of tyrosine hydroxylase activity is mediated by presynaptic dopamine receptors or by presynaptic allo-receptors. In addition, if a dopamine sensitive adenylate cyclase is involved in the regulation of neostriatal tyrosine hydroxylase activity it is relevant to know if this adenylate cyclase is linked to a D-1 and/or a D-2 receptor. At this point in time experimental data are not in favour of the presence of a D-1 receptor linked to an adenylate cyclase on the varicosities of dopaminergic neurons in the neostriatum. E.g. concentrations of dopamine agonists stimulating cyclic AMP formation inhibit tyrosine

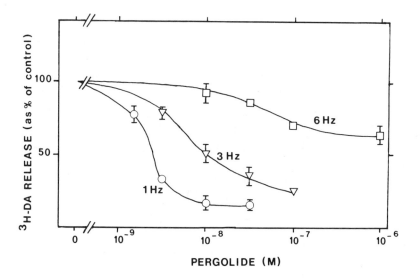

Figure 8.   The inhibition of electrically evoked ³H-dopamine release by pergolide, and its dependence on frequency of stimulation. (Reproduced with permission from Ref. 122. Copyright 1982, Pergamon Press, Ltd.)

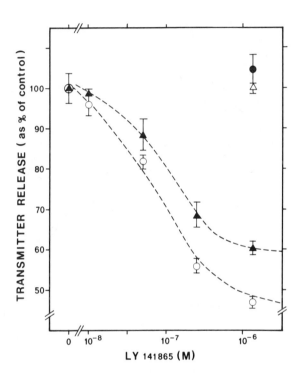

Figure 9. D-2 receptor agonist LY 141865 inhibits the K^+-stimulated release of [^{14}C]-acetylcholine (○) and [^3H]-dopamine (▲) from blocks of rat neostriatum. (−)-Sulpiride (1 μM) antagonized the effect of 1.25 μM LY 141865 on the [^{14}C]-acetylcholine release (●), and on the [^3H]-dopamine release (△). (Modified from Ref. 53.)

hydroxylase activity (see also ref. 134). It is tempting to
speculate that a D-2 like receptor, linked to an adenylate
cyclase, is located on the membranes of the dopaminergic
varicosities. Stimulation of this dopamine receptor would tend
to decrease cyclic AMP formation and consequently tyrosine
hydroxylase activity.

Differences between post- and pre-synaptic dopamine receptors

     The pre- and post-synaptic dopamine receptors are claimed
to be distinct entities on the basis of electrophysiological
and behavioral studies (50, 109). The demonstration of drugs
discriminating between the presynaptic and postsynaptic
dopamine receptor strengthens the claim that these two
receptors are distinct entities. The agonist N,N' dimethyl 6,7
dihydroxy 2 amino tetralin (TL-99) and N-n-propyl-(hydroxyphenyl)-piperidine (3PPP), and the antagonist, (-)-sulpiride,
are drugs claimed to discriminate between the pre- and postsynaptic dopamine receptors (135-138). Recent biochemical
studies have failed to substantiate these claims of anatomical
specificity. For example, although TL-99 can stimulate the
autoreceptor, it is also an effective agonist upon (both D-1
and D-2) postsynaptic dopamine receptors (139, 140, 141).
Similarly, in in vitro studies of dopamine release, 3-PPP does
not inhibit dopamine release (Sminia, Mulder and Stoof,
unpublished observations) or synaptosomal tyrosine hydroxylase
(142). In the case of (-)-sulpiride, different investigators
have claimed that it is specific for both postsynaptic dopamine
receptors (137) and presynaptic dopamine receptors (138). This
confused state suggests that attempts to distinguish The
Presynaptic Dopamine Receptor from The Postsynaptic Dopamine
Receptor may be misguided. The inability of several investigators to distinguish the autoreceptor inhibiting dopamine
release (a presynaptic dopamine receptor) from the dopamine
receptor inhibiting acetylcholine release (a postsynaptic D-2
dopamine receptor) raises the possibility that the autoreceptor
may ressemble some postsynaptic dopamine receptors (53, 132).

Concluding Remarks

     The aim of writing this chapter has been to present a short
review of the available data describing responses at the
cellular level which are consequences of dopamine receptor
stimulation in the neostriatum. By trying to identify and
characterize dopamine receptors in this way some promising
results have been obtained but many questions remain to be
answered.
     Apart from mediating a stimulation of cyclic AMP formation
no clear physiological functions has been discovered yet for
the (postsynaptic) D-1 receptor in the neostriatum or in other

parts of the CNS. Since physiological functions have been
reported for (pharmacologically) identical receptors in several
tissues outside the CNS, it is reasonable to presume a physiological function also within the CNS. Nevertheless it cannot be
excluded that e.g. (1) dopamine is not the physiological agonist
for this receptor in the mammalian CNS or that (2) this receptor
mediates trophic functions in the CNS. The recent discovery of
several drugs discriminating between the D-1 and D-2 receptors
will definitely help us to solve this problem in the near
future.

One of the CNS dopamine receptors identified and characterized most completely at the moment is the (postsynaptic) D-2
receptor mediating the inhibition of acetylcholine turnover and
release. The basis for this characterization traces back to
morphological studies presenting evidence for synaptic contacts
between dopaminergic varicosities and cholinergic neuronal
elements. Hopefully in the forthcoming years it will be possible
to collect morphological evidence for the presence of synaptic
contacts between dopaminergic varicosities and other than
cholinergic neuronal elements in the neostriatum or elsewhere.
Especially recently developed immunocytochemical approaches
could turn out to be extremely valuable for this experimental
approach.

Sofar (presynaptic) dopamine receptors mediating inhibition
of dopamine turnover and/or release seem to display features
resembling those of the D-2 receptor. However inconsistencies
reported could finally lead to a subdivision of D-2 receptors
in the near future or it will appear that some in vitro
experimental conditions have been too extreme, which by itself
would have induced changes in the pharmacological characteristic of the receptor. One obvious question emerging from this
review is whether all D-2 dopamine receptors in the neostriatum
are linked to an adenylate cyclase. With the methodology
presently available it will be hopefully only a matter of time
to answer this question.

Reviewing many papers for this chapter it was disturbing
to note that in a large number of studies drugs were used in
amounts or concentrations which would have justified publication
in toxicological journals rather than in pharmacological
journals. To use drugs in concentrations 1000-fold the
concentration needed to occupy a certain receptor does not give
much information about the mechanisms behind the response
elicited and the characteristics of the receptor under
investigation.

At this point in time the pharmacological characteristics
of the functionally different dopamine receptors, present in the
neostriatum, can (still) be described in terms of D-1 and D-2
receptors; there is (still) no need to assume the presence of
D-3 or D-4 receptors to explain the biochemical and physiological responses elicited by dopamine. It is conceivable

however that in the near future we may need a more differentiated subdivision to describe the characteristics of the dopamine receptors present in the neostriatum. Extension of these studies to other brain areas or other species may even intensify the need for such a subdivision.

Acknowledgments

I am grateful to Dr. Herman H. Harms for stimulating discussions and to Mr. Henk Dalecki for typing the manuscript.

Literature Cited

1. Watling, K.J.; Dowling, J.E.; Iversen, L.L. Nature 1979, 281, 578-580.
2. Brown, E.M.; Attie, M.F.; Reen, S.; Gardner, D.G.; Kebabian, J.W.; Aurbach, G.D. Mol. Pharmacol. 1980, 18, 335-340.
3. Munemura, M.; Cote, T.E.; Tsuruta, K.; Eskay, R.L.; Kebabian, J.W. Endocrinology 1980, 107, 1676-1683.
4. Seeman, P. Pharmacol. Rev. 1980, 32, 229-313.
5. Leysen, J.E.; Gommeren, W. J. Neurochem. 1981, 36, 201-219.
6. Andén, N.-E.; Carlsson, A.; Dahlström, A.; Fuxe, K.; Hillarp, N.-Å.; Larsson, K. Life Sci. 1964, 3, 523-530.
7. Andén, N.-E.; Fuxe, K.; Hamberger, B.; Hökfelt, T. Acta Physiol. Scand. 1966, 67, 306-312.
8. Falck, B.; Hillarp, N.-A.; Thieme, G.; Torp, A. J. Histochem. Cytochem. 1962, 10, 348-365.
9. Pickel, V.M.; Joh, T.H.; Reis, D.J. Adv. Biochem. Psychopharmacol. 1977, 16, 321-329.
10. Pickel, V.M.; Beckley, S.C.; Joh, T.H.; Reis, D.J. Brain Res. 1981, 225, 373-385.
11. Hassler, R. J. Neurol. Sci. 1978, 36, 187-224.
12. Fonnum, F.; Walaas, I. "The Neostriatum"; Divac, I.; Öberg, R.G.E., Eds.; Pergamon Press: Oxford, 1979; pp. 53-71.
13. Beinfeld, M.; Meyer, D.K.; Eskay, R.L.; Jensen, R.T.; Brownstein, M.J. Brain Res. 1981, 212, 51-57.
14. Nakamura, Y.; Hassler, R.; Kataoka, K.; Bak, I.J.; Kim, J.S. Folia psychiat. neurol. jap. 1976, 30, 185-194.
15. Hattori, T.; Singh, U.K.; McGeer, P.L. Brain Res. 1976, 102, 164-173.
16. Costa, E.; Cheney, D.L. "Interactions between putative neurotransmitters in the brain"; Garattini, S.; Pujol, J.F.; Samanin, R.S., Eds.; Raven Press: New York, 1978; pp. 23-38.
17. Guyenet, P.G.; Agid, Y.; Javoy, F.; Beaujouan, J.C.; Rossier, J.C.; Glowinski, J. Brain Res. 1975, 84, 227-244.
18. Ladinsky, H.; Consolo, S.; Bianchi, S.; Ghezzi, D.; Samanin, R. "Interactions between putative neurotransmitters in the brain"; Garattini, S.; Pujol, J.F.; Samanin, R., Eds.; Raven Press: New York, 1978; pp. 3-21.

19. Bartholini, G.; Lloyd, K.G.; Stadler, H. Adv. Neurol. 1974, 5, 11-17.
20. Euvrard, C.; Premont, J.; Oberlander, C.; Boissier, J.R.; and Bockaert, J. Naunyn-Schmiedeberg's Arch. Pharmacol. 1979, 309, 241-245.
21. Mulder, A.H. Progr. Brain Res. 1982, in press.
22. Scatton, B. J. Pharmacol. Exp. Ther. 1982, 220, 197-202.
23. Hertting, G.; Zumstein, A.; Jackisch, R.; Hoffman, I.; Starke, K. Naunyn-Schmiedeberg's Arch. Pharmacol. 1980, 315, 111-117.
24. Scatton, B.; Zivkovic, B.; Dedek, J. J. Pharmacol. Exp. Ther. 1980, 215, 494-499.
25. Lehmann, J.; Smith, R.V.; Langer, S.Z. Eur. J. Pharmacol. in press.
26. Stoof, J.C.; Thieme, R.E.; Vrijmoed-de Vries, M.C.; Mulder, A.H. Naunyn-Schmiedeberg's Arch. Pharmacol. 1979, 309, 119-124.
27. Stoof, J.C.; Horn, A.S.; Mulder, A.H. Brain Res. 1980, 196, 276-281.
28. Tsuruta, K.; Frey, E.A.; Grewe, C.W.; Cote, T.E.; Eskay, R.L.; Kebabian, J.W. Nature 1981, 292, 463-465.
29. Euvrard, C.; Ferland, L.; Dipaolo, T.; Beaulieu, M.; Labrie, F.; Oberlander, C.; Raynaud, J.P.; Boissier, J.R. Neuropharmacol. 1980, 19, 379-386.
30. Spano, P.F.; Stefanini, E.; Trabucchi, M.; Fresia, P. "Sulpiride and Other Benzamides"; Spano, P.F.; Trabucchi, M.; Corsini, G.U.; Gessa, G.L., Eds.; Italian Brain Research Foundation: Milan, 1979; pp. 11-31.
31. Scapagnini, U. "Sulpiride and Other Benzamides"; Spano, P.F.; Trabucchi, M.; Corsini, G.U.; Gessa, G.L., Eds.; Italian Brain Research Foundation: Milan, 1979; pp. 193-205.
32. Setler, P.E.; Sarau, H.M.; Zirkle,C.L.; Saunders, H.L. Eur. J. Pharmacol. 1978, 50, 419-430.
33. Watling, K.J.; Dowling, J.E. J. Neurochem. 1981, 36, 559-568.
34. Kaiser, C.; Dandridge, P.A.; Garvey, E.; Hahn, R.A.; Sarau, H.M.; Setler, P.E.; Bass, L.S.; Clardy, J. J. Med. Chem. 1982, 25, 697-703.
35. Hornykiewicz, O. "Handbook of Clinical Neurology, Vol. 29, Metabolic and Deficiency Diseases of the Nervous System"; Vinken, P.J.; Bruyn, G.W., Eds.; North Holland: Amsterdam, 1977; pp. 459-483.
36. Aquilonius, S.-M. "Parkinson's Disease: Current Progress, Problems and Management"; Rinne, U.K.; Klinger, M.; Stamm, G., Eds.; North Holland: Amsterdam, 1980; pp. 17-27.
37. Calne, D.B.; Williams, A.C.; Neophytides, A. Lancet 1978, 1, 735-738.
38. Marsden, C.D.; Parker, J.D. Lancet 1977, 1, 345-349.
39. Calne, D.B., Ed.; "Parkinsonism: Physiology, Pharmacology and Treatment"; Edward Arnold: London, 1970.

40. Fonnum, F.; Grofová, I.; Rinvik, E.; Storm-Mathisen, J.;
     Walberg, F. Brain Res. 1974, 71, 77-92.
41. Kim, J.S.; Bak, I.J.; Hassler, R.; Okada, Y. Exp. Brain
     Res. 1971, 14, 95-104.
42. McGeer, P.L.; Fibiger, H.C.; Maler, M.; Hattori, T.;
     McGeer E.G. Adv. Neurol. 1974, 3, 153-163.
43. Hassler, R.; Nitsch, C.; Lee, H.L. "Parkinson's Disease:
     Current Progress, Problems and Management"; Rinne, U.K.;
     Kingler, M.; Stamm, G., Eds.; North Holland: Amsterdam,
     1980; pp. 61-89.
44. Perez de La Mora, N.; Fuxe, K.; Hökfelt, T.; Ljungdahl, Å.
     Neurosci. Lett. 1975, 1, 109-114.
45. Kim, J.S.; Hassler, R. Brain Res. 1975, 88, 150-153.
46. Marco, E.; Mao, C.C.; Cheney, D.L.; Revuelta, A.; Costa, E.
     Nature 1976, 264, 363-365.
47. Gale, K.; Casu, M. Mol. Cell. Biochem. 1981, 39, 369-405.
48. Van der Heyden, J.A.M.; Venema, K.; Korf, J. J. Neurochem.
     1980, 34, 1338-1341.
49. Van der Heyden, J.A.M.; Venema, K.; Korf, J. J. Neurochem.
     1980, 34, 119-125.
50. Skirboll, L.R.; Grace, A.A.; Bunney, B.S. Science 1979,
     206, 80-82.
51. Leysen, J. Adv. Human Psychopharmacol. 1982, 3, in press.
52. Pycock, C.J.; Horton, R.W.; Carter, C.J. Adv. Biochem.
     Psychopharm. 1978, 19, 323-346.
53. Stoof, J.C.; de Boer, Th.; Sminia, P.; Mulder, A.H.
     Eur. J. Pharmacol. 1982, in press.
54. Brase, D.A. J. Pharm. Pharmacol. 1980, 32, 432-433.
55. Reubi, J.-C.; Iversen, L.L.; Jessel, T.M. Nature 1977, 268,
     652-654.
56. Arbilla, S.; Kamal, L.A.; Langer, S.Z. Br. J. Pharmacol.
     1981, 74, 389-397.
57. Webster, K.E. J. Anat. (Lond.) 1961, 95, 532-545.
58. Spencer, H.J. Brain Res. 1976, 102, 91-101.
59. McGeer, P.L.; McGeer, E.G.; Scherer, U.; Singh, K.
     Brain Res. 1977, 128, 369-373.
60. Divac, I.; Fonnum, F.; Storm-Mathisen, J. Nature 1977, 266,
     377-378.
61. Godukhin, O.V.; Zharikova, A.D.; Novoselov, V.I.
     Neuroscience 1980, 5, 2151-2154.
62. Rowlands, G.J.; Roberts, P.J. Exp. Brain Res. 1980, 39,
     239-240.
63. Schwarcz, R.; Creese, I.; Coyle, J.T.; Snyder, S.H. Nature
     1978, 271, 766-768.
64. Rowlands, G.J.; Roberts, P.J. Eur. J. Pharmacol. 1980, 62,
     241-242.
65. Mitchell, P.R.; Doggett, N.S. Life Sci. 1980, 26,
     2073-2081.
66. Fuxe, K.; Jonsson, G. Adv. Biochem. Psychopharm. 1974, 10,
     1-12.

67. Lorens, S.A.; Guldberg, H.C. Brain Res. 1974, 78, 45-45.
68. Geyer, M.A.; Puerto, A.; Dowsey, W.J.; Knapp, S.; Bullard W.P.; Mandell, A.J. Brain Res. 1976, 106, 241-256.
69. Dray, A.; Gonye, T.J.; Oakley, N.R.; Tanner, T. Brain Res. 1976, 113, 45-57.
70. Pasik, P.; Pasik, T.; Di Figlia, M. "The Neostriatum"; Divac, I.; Öberg, R.G.E., Eds.; Pergamon Press: Oxford, 1979; pp. 5-36.
71. Hery, F.; Soubrie, P.; Bourgoin, S.; Motastrue, J.L.; Artand, F.; Glowinski, J. Brain Res. 1980, 193, 143-151.
72. Reubi, J.C.; Emson, P.C.; Jessell, T.M.; Iversen, L.L. Naunyn-Schmiedeberg's Arch. Pharmacol. 1978, 304, 271-275.
73. Hökfelt, T.; Johansson, O.; Ljungdahl, Å.; Lundberg, J.M.; Schultzberg, M. Nature 1980, 284, 515-521.
74. Ljungdahl, Å.; Hökfelt, T.; Nilsson, G. Neuroscience 1978, 3, 861-943.
75. Hanson, G.R.; Alphs, L.; Wolf, W.; Levine, R.; Lovenberg, W. J. Pharmacol. Exp. Ther. 1981, 218, 568-574.
76. Pettibone, D.J.; Wurtman, R.J. Brain Res. 1980, 186, 409-419.
77. Hanson, G.; Alphs, L.; Pradhan, S.; Lovenberg, W. Neuropharmacol. 1981, 20, 541-548.
78. Hökfelt, T.; Elde, R.; Johansson, O.; Terenius, L.; Stein, L. Neurosci. Lett. 1977, 5, 25-31.
79. Klaff, L.J.; Hudson, A.M.; Paul, M.; Millar, R.P. Peptides 1982, 1, 155-161.
80. Williams, J.A. Biomed. Res. 1982, 3, 107-121.
81. Osborne, H.; Herz, A. Naunyn-Schmiedeberg's Arch. Pharmacol. 1980, 310, 203-209.
82. Krueger, B.K.; Forn, J.; Walters, J.R.; Roth, R.H.; Greengard, P. Mol. Pharmacol. 1976, 12, 639-648.
83. McGeer, E.G.; Innanen, V.T.; McGeer, P.L. Brain Res. 1976, 118, 356-359.
84. DiChiara, G.; Porceddu, M.L.; Spano, P.F.; Gessa, G.L. Brain Res. 1977, 130, 374-382.
85. Clement-Cormier, Y.C.; Parrish, R.G.; Petzold, G.L.; Kebabian, J.W.; Greengard, P. J. Neurochem. 1975, 25, 143-149.
86. Spano, P.F.; DiChiara, G.; Tonon, G.C.; Trabucchi, M. J. Neurochem. 1976, 27, 1565-1568.
87. Kebabian, J.W.; Saavedra, J.M. Science 1976, 193, 683-685.
88. Gale, K.; Guidotti, A.; Costa, E. Science 1977, 195, 503-505.
89. Miller, R.J.; McDermed, J. "The Neurobiology of dopamine"; Horn, A.S.; Korf, J.; Westerink, B.H.C., Eds.; Academic Press: London, 1979; pp. 159-178.
90. Iversen, L.L. Science 1975, 188, 1084-1089.
91. DeCamilli, P.; Macconi, D.; Spada, A. Nature 1979, 278, 252-254.
92. Giannattasio, G.; DeFerrari, M.E.; Spada, A. Life Sci. 1981, 28, 1605-1612.

93. Swennen, L.; Denef, C. Endocrinology 1982, in press.
94. Onali, P.; Schwartz, J.P.; Costa, E. Proc. natl. Acad.
    Sci. U.S.A. 1981, 78, 6531-6534.
95. Stoof, J.C.; Kebabian, J.W. Nature 1981, 294, 366-368.
96. Stoof, J.C.; Kebabian, J.W. Brain Res., 1982, in press.
97. Goldstein, M.; Freedman, L.S.; Backstrom, T.J. J. Pharm.
    Pharmacol. 1970, 22, 715-717.
98. Goldstein, M.; Anagnoste, B.; Shirron, C. J. Pharm.
    Pharmacol. 1973, 25, 348-351.
99. Christiansen, J.; Squires, R.F. J. Pharm. Pharmacol. 1974,
    26, 367-369.
100. Iversen, L.L.; Roganski, M.; Miller, R.J. Mol. Pharmacol.
     1976, 12, 251-262.
101. Westfall, T.C.; Besson, M.-J.; Giorguieff, M.-F.;
     Glowinski, J. Naunyn-Schmiedeberg's Arch. Pharmacol. 1976,
     292, 279-287.
102. Cerrito, F.; Mauro, G.; Raiteri, M. "Apomorphine and Other
     Dopaminomimetics"; Gessa, G.L.; Corsini, G.U., Eds.;
     Raven Press: New York, 1981; pp. 123-132.
103. Andén, N.-E.; Roos, B.-E.; Werdinius, B. Life Sci. 1964,
     3, 149-158.
104. Burkard, W.P.; Grey, K.F.; Pletscher, A. Nature 1967, 213,
     732-733.
105. Scatton, B.; Bischoff, S.; Dedek, J.; Korf, J. Eur. J.
     Pharmacol. 1977, 44, 287-292.
106. Westerink, B.H.C.; Lejeune, B.; Korf, J.; van Praag, H.M.
     Eur. J. Pharmacol. 1977, 42, 179-190.
107. Westerink, B.H.C. Eur. J. Pharmacol. 1979, 56, 313-322.
108. Hofmann, M.; Battaini, F.; Tonon, G.; Trabucchi, M.;
     Spano, P.F. Eur. J. Pharmacol. 1979, 56, 15-20.
109. Roth, R.H. Comm. Psychopharmacol. 1979, 3, 429-445.
110. Bunney, B.S.; Walters, J.R.; Roth, R.H.; Aghajanian, G.K.
     J. Pharmacol. Exp. Ther. 1973, 185, 560-571.
111. Bunney, B.S.; Aghajanian, G.K.; Roth, R.H. Nature 1973,
     245, 123-125.
112. Guyenet, P.G.; Aghajanian, G.K. Brain Res. 1978, 150,69-84.
113. Bannon, M.J.; Bunney, E.B.; Zigun, J.R.; Skirboll, L.R.;
     Roth, R.H. Naunyn-Schmiedeberg's Arch. Pharmacol. 1980,
     312, 161-165.
114. Starke, K. Ann. Rev. Pharmacol. Toxicol. 1981, 21, 7-30.
115. Langer, S.Z. Pharmacol. Rev. 1980, 32, 337-362.
116. Farnebo, L.O.; Hamberger, B. Acta physiol. scand. Suppl.
     1971, 371, 35-44.
117. Plotsky, P.M.; Wightman, R.M.; Chey, W.; Adams, R.N.
     Science 1977, 197, 904-906.
118. Starke, K.; Reimann,W.; Zumstein, A.; Hertting, G.
     Naunyn-Schmiedeberg's Arch. Pharmacol. 1978, 305, 27-36.
119. Dismukes, R.K.; Mulder, A.H. Naunyn-Schmiedeberg's Arch.
     Pharmacol. 1977, 297, 23-29.
120. Miller, J.C.; Friedhoff, J.C. Biochem. Pharmacol. 1979,
     28, 688-690.

121. Raiteri, M.; Cervoni, A.M.; del Carmine, R. Nature 1978,
      274, 706-708.
122. Lehmann, J.; Langer, S.Z."Advances in Dopamine Research"
      Pergamon Press, in press.
123. Reimann, W.; Zumstein, A.; Jackisch, R.; Starke, K.;
      Hertting, G. Naunyn-Schmiedeberg's Arch. Pharmacol. 1979,
      306, 53-60.
124. Kamal, L.A.; Arbilla, S.; Langer, S.Z. J. Pharmacol. Exp.
      Ther. 1981, 216, 592-598.
125. Jackisch, R.; Zumstein, A.; Hertting, G.; Starke, K.
      Naunyn-Schmiedeberg's Arch. Pharmacol. 1980, 314, 129-133.
126. Arbilla, S.; Langer, S.Z. Eur. J. Pharmacol. 1981, 76,
      345-351.
127. Rabey, J.M.; Passeltiner, P.; Markey, K.; Asano, T.;
      Goldstein, M. Brain Res. 1981, 225, 347-356.
128. Scatton, B. Eur. J. Pharmacol. 1977, 46, 363-369.
129. Westerink, B.H.C. "The Neurobiology of dopamine"; Horn,
      A.S.; Korf, J.; Westerink, B.H.C., Eds.; Academic Press:
      London, 1979; pp. 255-291.
130. Jenner, P.; Marsden, C.D. "Sulpiride and other Benzamides";
      Spano, P.F.; Trabucchi, M.; Corsini, G.U.; Gessa, G.L.,
      Eds.; Italian Brain Research Foundation: Milan, 1979;
      pp. 119-147.
131. Marek, K.L.; Roth, R.H. Eur. J. Pharmacol. 1980, 62,
      137-146.
132. Helmreich, I.; Reimann, W.; Hertting, G.; Starke, K.
      Neuroscience 1982, 7, 1559-1566.
133. Roth, R.H.; Walters, J.R.; Murrin, L.C.; Morgenroth, V.H.
      "Pre- and Postsynaptic Receptors"; Usdin, E.; Bunney, W.E.,
      Eds.; Marcel Dekker: New York, 1975; pp. 5-48.
134. Kehr, W.; Debus, G. "Presynaptic Receptors": Langer, S.Z.;
      Starke, K.; Dubocovich, M.L., Eds.; Pergamon Press: Oxford,
      1979, pp. 199-206.
135. Goodale, D.B.; Rusterholz, D.B.; Long, J.P.; Flynn, J.R.;
      Walsh, B.; Cannon, J.G.; Lee, T. Science 1980, 210,
      1141-1143.
136. Hjörth, S.; Carlsson, A.; Wikström, H.; Lindberg, P.;
      Sanchez, D.; Hacksell, U.; Arvidsson, L.E.; Svensson, U.;
      Nilsson, J.L.G. Life Sci. 1981, 28, 1225-1238.
137. Ålander, T.; Andén, N.-E.; Grabowska-Andén, M. Naunyn-
      Schmiedeberg's Arch. Pharmacol. 1980, 312, 145-150.
138. Puech, A.J.; Simon, P.; Boissier, J.R. Eur. J. Pharmacol.
      1978, 50, 291-300.
139. Horn, A.S.; de Vries, J.; Dijkstra, D.; Mulder, A.H.
      Eur. J. Pharmacol. 1982, in press.
140. Martin, G.E.; Haubrich, D.R.; Williams, M. Eur. J.
      Pharmacol. 1981, 76, 15-23.
141. Watling, K.J.; Williams, M. Eur. J. Pharmacol. 1982, 77,
      321-329.
142. Haubrich, D.R.; Pflueger, A.B. Mol. Pharmacol. 1982, 21,
      114-120.

RECEIVED November 4, 1982

Potential Therapeutic Uses of Dopamine Receptor Agonists and Antagonists

DONALD B. CALNE and T. ANDREO LARSEN

University of British Columbia, Health Sciences Centre Hospital, Division of Neurology, Vancouver, British Columbia, Canada

The various types of dopamine agonists are reviewed, and their applications are discussed in neurological, endocrinological and cardiovascular disease. The major indications for dopamine agonists are Parkinson's Disease, hyperprolactinemia, acromegaly, certain pituitary tumors, and shock. Dopamine antagonists are then considered; these drugs are employed predominately for the treatment of psychiatric disease, notably schizophrenia. More drugs are required with selective actions on the different categories of dopamine receptor; when these are developed it should be possible to improve the therapeutic index of treatment for the above diseases and extend perhaps the application of these agents to other disorders.

It is less than 25 years since dopamine (DA) became established as a neurotransmitter, rather than simply existing as a precursor for norepinephrine (1). Yet DA has overtaken acetylcholine and norepinephrine as the most extensively investigated neurotransmitter in the nervous system, and it has now been found to have hormonal functions (in the portal circulation of the pituitary), in addition to its activity as a humerol agent for communication between nerve cells. Analyses of the receptor mechanisms for dopamine have proved more difficult than comparable studies for acetylcholine and norepinephrine, partly because of the limited range of artificial agonists and antagonists. Nevertheless, at least two types of DA receptor have been characterized,(2) one associated with an adenylate cyclase (D-1 receptors) and another independent of this enzyme (D-2 receptors). Other categories of DA receptor may exist, but the evidence for defining these is less complete (3).
    The clinical impact of the surge in new knowledge on DA has been substantial, but it is salutory to recognize that the largest

0097–6156/83/0224–0147$06.00/0
© 1983 American Chemical Society

single application of drugs that modify DA function - the use of
DA antagonists in schizophrenia - derives from pragmatic
observations which preceded even the first speculation that DA is
a neurotransmitter. In spite of this humbling example of lack of
scientific rationale for a major therapeutic advance, it is
reasonable to claim that the logical application of
pharmacological principles has led to important developments in
the treatment of a wide range of human diseases.

Dopamine Agonists

     The main agents to have been employed to increase
dopaminergic effects are:-
1.   DA itself.
2.   The precursor of DA, L-DOPA, whose actions differ from
     dopamine by virtue of pharmacokinetic factors - unlike DA,
     L-DOPA can cross the blood brain barrier with reasonable
     ease.
3.   Amphetamine, a drug that acts by releasing DA from nerve
     terminals.
4.   Apomorphine and its derivatives. These drugs are of interest
     as the first articial, direct acting agonists. Their use is
     limited by rather prominent unwanted actions.
5.   Ergot derivatives. The first and most widely prescribed ergot
     derivative with dopaminomimetic properties is bromocriptine.
     Congeners of bromocriptine that are currently undergoing
     clinical evaluation include lysuride, pergolide, and CU
     32-085.
     Before reviewing the clinical conditions in which DA agonists
have a therapeutic role, it is appropriate to mention occasional
paradoxes that occur, where DA agonists appear to be eliciting
responses that are generally regarded as the opposite of those
expected. For example, very high doses of L-DOPA have been
reported, on rare occasions, to induce an exacerbation of
Parkinson's disease; this is thought to be analogous to the way in
which high concentrations of cholinergic agents can lead to a
depolarization block at cholinergic synapses. Similarly, some
artifical agonists of DA may act as DA antagonists at high tissue
levels, either by the above mechanism, or by behaving as "partial
agonists" (which stimulate receptors at low concentrations but
inhibit them at high tissue levels). Finally, very low
concentrations of certain dopaminomimetics may selectively
activate presynaptic receptors, to elicit a homeostatic, "negative
feedback" effect of decreasing the release of DA from nerve
endings; this mechanism has been postulated to account for the
occasional improvement in choreatic and dystonic movement
disorders, reported, for example, with low doses of apomorphine
and bromocriptine (4,5,6).
     All the DA agonists employed clinically activate D-2
receptors, but none are entirely selective. Some also stimulate

D-1 receptors; these include DA, L-DOPA, amphetamines, apomorphines and certain ergot derivatives such as pergolide. Other D-2 agonists ergot derivatives inhibit the D-1 receptor; these include bromocriptine, lysuride, and CU32-085. The commoner adverse reactions induced by DA agonists comprise nausea, hypotension, dyskinesia, hallucinations and delusions.

Neurological Disorders. The major application for DA agonists in neurology is the treatment of Parkinson's disease (7). The rationale for this is a logical example of therapeutic science. DA is a neurotransmitter that normally exists in very high concentrations in the striatum of the brain, but in Parkinson's disease the nerve cells that produce and release DA undergo degeneration. This situation can be markedly alleviated by oral administration of DA agonists. By comparing the relative action of different DA agonists, it seems that activation of the D-2 receptor is particularly important for the achievement of a therapeutic response in Parkinson's disease.

While Parkinson's disease represents by far the commonest neurological disorder amenable to treatment by DA agonists, the paradoxical action of these drugs has occasionally been reported to elicit favorable responses in Huntington's disease (4,6), dystonia musculorum deformans (5) and certain segmental and focal dystonias.

Endocrinological Disorders. The discovery that DA is a potent prolactin inhibitory factor in the portal circulation of the pituitary is the basis for the use of DA agonists to suppress unwanted normal puerperal lactation, and in treating pathological hyperprolactinaemia together with its associated anovular infertility.

A further endocrinological application for dopamine agonists is the treatment of acromegaly. Normally, dopaminomimetics stimulate the release of growth hormone from the pituitary, but in acromegaly there is an unexplained reversal of the response such that DA agonists suppress the abnormal elevation of growth hormone. These agents represent the first medical form of treatment for acromegaly.

Finally, another serendipitous discovery is having a major impact on endocrinology. It has been found that the same long acting, artificial DA agonists (ergot derivatives) that have been employed to treat hyperprolactinaemia and acromegaly induce atrophy of certain pituitary tumors, in particular those that secrete prolactin. The mechanism of this antitumor effect is unknown, but here again a new medical treament has become available for what was previously regarded as a disease that required either surgical or X-ray therapy.

One curious aspect of the action of DA agonists on pituitary tumors is the remarkable speed of their effect. Dramatic changes in tumor size, with clinical concomitants such as shrinking of

visual field defects, can occur within 10 days of starting
treatment (8).

Cardiovascular Disorders. DA itself is a useful agonist
where intravenous therapy is required for immediate effects
limited to the periphery (i.e. outside the blood brain barrier).
It has therefore been employed as a pharmacolgical agent in shock
- a condition in which there is no disease involving dopaminergic
mechanisms, but where the inotropic actions on the heart and
vasodilator effects on the kidney can reverse the potentially
lethal consequences of profound arterial hypotension, whatever its
cause. This subject is discussed in detail elsewhere in this
symposium.

Other applications for dopamine agonists. The traditional
use of apomorphine as an emetic stems from its dopaminomimetic
actions in the brainstem. This mechanism of action has been
identified retrospectively, since apomorphine was employed to
induce vomiting long before it was known to be a DA agonist.
Other areas of interest as potential applications for DA
agonists include alleviation of premenstural tension, correction
of normoprolactinaemic infertility, the treatment of mastalgia,
impotence, Cushing's syndrome, and hypertension (8). The use of
DA agonists in peptic ulcer is a very recent application discussed
further in another section of this book (Chapter 8). The validity
of the claims for therapeutic efficacy in these conditions remains
unproven, and the mechanisms by which DA agonists might elicit
these benefical effects are ill defined. More clinical
information is needed before any conclusions can be drawn on the
possible value of dopaminomimetics in these settings.

Dopamine Antagonists

A variety of drugs that block DA receptors are available for
clinical use, and even more for experimental purposes. These
drugs, also referred to as neuroleptics, include phenothiazines
(e.g., chlorpromazine), thioxanthenes (e.g., chlorprothixene),
butyrophenones (e.g., haloperidol), diphenylbutylpiperidines
(e.g., pimozide), and dibenzodiazepines (e.g., clozapine). The
major medical applications for these drugs are in the treatment of
severe psychiatric illnesses, certain movement disorders, emesis
and intractable hiccough.
The differential clinical actions of DA blockers on the DA
receptor subtypes have not been defined with precision. Most
neuroleptics appear to act at both D-1 and D-2 receptors. Some
differences exist, however. The thioxanthenes bind to sites
related to both DA receptor subtypes, but the butyrophenones seem
to prefer sites assocaited with D-2 receptors, binding only weakly
to those identified with D-1 receptors. Sulpiride, molindone and
metoclopramide are relatively selective D-2 antagonists.

It appears, therefore, that D-1 blockade is not relevant to the antipsychotic effect or suppression of hyperkinetic movement disorders.

Many earlier reports considered extrapyramidal side effects unavoidable when treating patients with neuroleptics. There appeared to be a parallel   between antipsychotic action and the incidence of unwanted neurological effects. However, the development of newer neuroleptics has changed this view. Drugs like clozapine have a high antipsychotic potency and yet produce few neurological problems. It has therefore been proposed that the DA receptors involved in the beneficial actions of neuroleptics in the treatment of psychiatric disorders are situated in mesolimbic areas, such as the nucleus accumbens, whereas the extrapyramidal effects are mediated by striatal receptors.

By blocking the striatal DA receptors, the neuroleptic drugs may cause several unwanted neurological reactions; these comprise acute dystonic reactions, tardive dyskinesia and akathisia (motor restlessness). The most common and the most difficult management problem is tardive dyskinesia, a late reaction appearing after months or years of treatment. The proposed mechanism is a pharmacologically induced "denervation hypersensitivity", although other explanations have been proposed e.g. overactivity in the noradrenergic systems (9). There is no satisfactory treatment available for tardive dyskinesia. The piperazine group of phenothiazine drugs and haloperidol are particularly prominent in causing this problem; there is less likelihood with thioridazine, clozapine and sulpiride (10).

Psychiatric Disorders. The main indications for DA antagonists are the treatment of adult schizophrenia, and childhood psychosis. Haloperidol and chlorpromazine are the most frequently employed.

In addition to acute and chronic schizophrenia, the neuroleptics are sometimes used in the management of mania, delirium, and severe agitation, whatever the cause of these symptom complexes. It must be noted that unlike parkinsonism, where a definite dysfunction in the DA system has been established, for schizophrenia and other psychiatric diseases, no unequivocal evidence has yet been presented to prove that there is a disturbance of the DA system (e.g., dopaminergic overactivity or receptor hypersensitivity). In untreated schizophrenics the production of DA metabolites is normal. Conflicting results have been obtained in studies of the DA receptors in schizophrenics (11,12,13), but in the case of patients who have not received neuroleptics, the receptor density and affinity appear to be normal (13). The "dopamine hypothesis" in these disorders derives from the beneficial effects of drugs that block DA receptors.

Neurological Disorders. Huntington's chorea is a hereditary

disease characterized by choreatic involuntary movements and mental deterioration. There is gross atrophy of the corpus striatum, with additional neural degeneration in the frontal cerebral cortex. The choreatic movement disorder responds to phenothiazines and butyrophenones. There is, however, no treatment for the mental decline.

Improvement has been reported following the use of haloperidol and chlorpromazine in other choreatic movement disorders including senile chorea, Sydenham's chorea and the chorea asociated with hyperthyroidism (14,15).

The syndrome of Gilles de la Tourette is a rare childhood illness characterized by multiple, chronic tics, grunts and vocalizations that are frequently obscene (coprolalia). Neuroleptics alleviate this clinical picture. Although most of the therapeutic experience in this disorder has been gained with haloperidol, phenothiazine neuroleptics have also been shown to be efficaceous (16).

Emesis. DA receptor blocking agents suppress nausea and vomiting. Phenothiazines, such as chlorpromazine and prochlorperazine, are usually employed. Their action is probably mediated via blockade of DA receptors in the chemoreceptor trigger zone of the medulla oblongata.

Hiccough. Chlorpromazine and prochlorperazine have been used in the treatment of intractable hiccough (10). Their mechanism of action is unknown.

Conclusions

Manipulation of the DA receptors is employed in therapeutics for neurological, endocrinological, psychiatric and cardiovascular disorders. When further knowledge of the receptors is gained, it is probable that drugs will be developed with more selective actions on various subtypes in diverse locations. These pharmacological refinements should result in an increase in the therapeutic index of treatment - an improvement in the balance between wanted and unwanted effects.

Literature Cited

1. Carlsson, A. Pharmacol. Rev. 1959,11,490.
2. Kebabian, J.W.; Calne, D.B. Nature 1979,277,93-6.
3. Seeman, P. Pharmacol. Rev. 1980,32.229-313
4. Frattola, L.; Albizzati, M.G.; Bassi, S.; Trabucchi, M. Dyskinesia, dystonia and dopaminergic system: effect of apomorphine and ergot alkaloias. In "Apomorphine and Other Dopaminomimetics".Vol.2, ed. Corsini, G.U. and Gessa, G.L. Raven Press: New York, 1981, 145-51.
5. Stahl, S.M.; Berger, P.A. Lancet 1981,2,745.

6. Tolosa, E.S.; Sparber, S.B. Life Sci. 1975, 15, (7) 1371-80.

7. Calne, D.B. "Therapeutics in Neurology", Blackwell Scientific Publications: Oxford, 1980.

8. Thorner, M.O.; Fluckiger, E.; Calne, D.B. "Bromocriptine: A Clinical and Pharmacological Review". Raven Press: New York, 1980

9. Jeste, D.V.; Wyatt, R.J. J. Clin. Psychiat. 1981,42,455-7.

10. Baldessarini, R.J. Drugs and the treatment of psychiatric disorders. In Goodman Gilman A, Goodman L.S. and Gilman A (eds.) "The Pharmacological Basis of Therapeutics". VIth ed. Macmillan, New York, 1980, 391-47.

11. Lee, T.; Seeman, P. Am. J. Psychiat. 1980,137,191-7.

12. Seeman, P. Lancet 1981,1,1103.

13. Mackay, A.V.P.; Bird, O.; Bird, E.D.; Spokes, E.G.; Rosser,M; Iversen,L.L.; Creese,I; Snyder,S.H. Lancet 1980,2,915-6

14. Shenker, D.M.; Grossman, H.; Klawans, H.L. Dev. Med. Child. Neurol. 1973, 15, 19-24.

15. Klawans, H.L.; Shenker, D.M.;Weiner, W.J. Observations on the dopaminergic nature of chorea. In "Advances in Neurology." Barbeau,A.;Chase T.N.; Paulson G.W.(eds.) Raven Press, New York, 1973, 547-9.

16. Levy, B.S.; Ascher, E. J. Neurol. Ment. Dis. 1968, 146,36-40.

RECEIVED February 16, 1983

Commentary: Potential Therapeutic Uses of Dopamine Receptor Agonists and Antagonists

MICHAEL O. THORNER

University of Virginia Medical Center, Charlottesville, VA 22908

This chapter reviews the clinical applications of the
dopamine system in man. It gives a balanced view of the
subject. There are only a few areas where I feel qualified to
review the material presented.

1. The classification of dopamine receptors. Although
Kebabian and Calne (1) performed an excellent service to the
scientific community in proposing the D-1 and D-2 receptor
hypothesis, this remains to be proven by isolating and
characterizing these receptors. These have been excellently
reviewed by Cronin (2). Indeed the statement that the "D"
receptors of the anterior pituitary are independent of adenylate
cyclase is incorrect. We (3) and others (4,5) have shown a
consistant and specific inhibitory relationship of the anterior
pituitary dopamine receptor and adenylate cyclase.

2. I believe it is dangerous to claim that dopamine agonist
drugs lead to atrophy of some prolactin-secreting pituitary
tumors. Atrophy implies that the tumors may ultimately
disappear. In our own experience we have shown that as soon as
treatment is withdrawn, prolactin levels rise again, and the
tumors increase in size again, and visual field defects may
recur. In a study performed with Drs. Tindall (Emory University)
and Kovacs and Horvath (Toronto University) we have been able to
show marked changes in the morphology of prolactinoma cells
during bromocriptine therapy (6). The cells lose the
characteristics of active secretion and there is a marked
reduction in the size of the individual cells. When the
treatment is withdrawn the changes are reversed, thus the
dopamine agonists suppress the synthesis and secretion of
prolactin for as long as they are given; this is associated with
morphological changes in the cells which ultimately lead to a
reduction in tumor size. However, this is not a permanent
effect and when therapy is withdrawn as long as one year after
initiation of therapy the clinical picture reverses, often
within days to that seen prior to treatment with an increase in

0097–6156/83/0224–0154$06.00/0

prolactin levels, increase in tumor size, and recurrence of compressive symptoms if they were initially present. Thus this therapy is useful providing it is continued probably life long.

3. The discussion of the role of bromocriptine in acromegaly confirms my own personal belief that it is useful adjunctive treatment but in contrast to prolactinomas, acromegaly should probably be treated primarily with pituitary surgery and if that fails, with irradiation and/or medical therapy with dopamine agonist drugs. It should be stressed that there is some dispute in literature as to the efficacy of dopamine agonists in the treatment of acromegaly (7,8). In general it is agreed that bromocriptine therapy gives excellent clinical results but the mechanism of this improvement is unclear. Growth hormone levels are rarely suppressed to normal and somatomedin C levels are only variably affected (9). The clinical response is much greater than what one would expect judging from the biochemical data; however, this is not a new observation related to bromocriptine. Similar observations have been made after pituitary surgery and radiation therapy.

4. One of the drugs used to suppress nausea, metoclopramide, also a dopamine receptor blocking drug, is now becoming increasingly used particularly in nausea associated with the administration of chemotherapeutic agents.

Literature Cited

1. Kebabian, J. W.; Calne, D. B. Nature (London) 1979, 277, 93.
2. Cronin, M. J. "Neuroendocrine Perspectives"; Elsevier Biomedical Press: New York, 1982; p 169.
3. Cronin, M. J.; Thorner, M. O. J. Cyclic Nucleotide Res., in press.
4. Swennen, L.; Denef, C. Endocrinology 1982, 111, 398.
5. Giannattasio, G.; DeFerrari, M. E.; Spada, A. Life Sci. 1981, 28, 1605.
6. Tindall, G.; Kovacs, K.; Horvath, E.; Thorner, M. O. J. Clin. Endocrinol. Metab. 1982, 55, 1178.
7. Lindholm, J.; Riishede, J.; Vestergaard, S.; Hummer, L.; Faber, O.; Hagen, C. N. Engl. J. Med. 1981, 304, 1450.
8. Thorner, M. O.; Besser, G. M.; Wass, J. A. H.; Liuzzi, A.; Hall R.; Muller, E. E.; Chiodini, P. G. N. Engl. J. Med. 1981, 305, 1092.
9. Wass, J. A. H.; Clemmons, D. R.; Underwood, L. E.; Barrow, I.; Besser, G. M.; Van Wyk, J. J. Clin. Endocr. 1982, 17, 369.

RECEIVED January 27, 1983

Dopaminergic Benzazepines with Divergent Cardiovascular Profiles

JOSEPH WEINSTOCK, JAMES W. WILSON, DAVID L. LADD,
and MARTIN BRENNER

Smith Kline & French Laboratories, Medicinal Chemistry Department,
Philadelphia, PA 19101

DENNIS M. ACKERMAN, ALAN L. BLUMBERG, RICHARD A. HAHN,
J. PAUL HIEBLE, HENRY M. SARAU, and VIRGIL D. WIEBELHAUS

Smith Kline & French Laboratories, Pharmacology Department,
Philadelphia, PA 19101

The pharmacological profile of the benzazepines I-IV
suggests that a range of dopaminergic activities result from activation of similar, but not identical,
receptors. SK&F 38393 (I) has both central and peripheral agonist activity, with the latter including
modest renal vasodilator and substantial diuretic
activity. SK&F 82526 (II) does not pass the blood
brain barrier, has potent renal vasodilator activity, and modest diuretic and antihypertensive activity. SK&F 87516 (III) has renal vasodilator potency comparable to II, but also has substantial diuretic and bradycardiac effects. SK&F 85174 (IV)
has renal vasodilator and substantial bradycardiac
effects. The bradycardiac effect has been characterized as arising from presynaptic dopamine agonist activity, while I-IV are all at least as potent
as dopamine in stimulating rat striatal adenylate
cyclase.

The elegant systems analysis of the mechanisms of blood pressure control by Guyton (1) suggested that impaired renal blood
flow is an important determinant of hypertension. Concurrent
investigations by Goldberg (2) on the renal effects of dopamine
showed that low doses of dopamine increased renal blood flow and
reduced renal vascular resistance. This work was confirmed and
extended by investigators in our laboratories (3,4,5) who found
that these responses were not inhibited by α- and β-receptor antagonists (phenoxybenzamine and propranolol), but were antagonized by bulbocapnine and metoclopramide. These effects have
been characterized as post-synaptic (6) and appear to be adenylate cyclase linked (7) or D-1 receptor mediated (8). These
intriguing observations led us to initiate a search for peripherally acting dopamine agonists which would be useful as therapeutic agents for the treatment of hypertension.

The required properties of such an agent included (1) selectivity for peripheral vascular dopaminergic receptors versus α- and β-adrenergic receptors which could mediate pressor and cardiac effects, (2) absence of central dopaminergic and emetic effects, and (3) potent oral renal vasodilator effects. Dopamine has been associated with diuresis and natriuresis. Possible mechanisms include a direct tubular effect on sodium transport, indirect effects produced by changes in total or regional renal blood flow, or effects resulting from a dopamine induced decrease in aldosterone release from the adrenal (9). Since diuretics play a key role in antihypertensive therapy, the addition of a natriuretic/diuretic component to the renal vasodilator profile would be valuable and appeared to be feasible.

Another important property of dopamine is its ability to inhibit sympathetic nerve function by interacting with presynaptic dopaminergic receptors to decrease norepinephrine release (10). These receptors are not adenylate cyclase coupled and have been classified as D-2 (8). Activation of cardiac presynaptic dopamine receptors causes bradycardia, and of vascular presynaptic dopamine receptors passive vasodilation, the magnitude of which will depend on the contribution of adrenergic activity to maintaining heart rate and vascular smooth muscle tone (11,12). Since arterial blood pressure and cardiac rate are important determinants of myocardial work and oxygen consumption, a presynaptic cardioactive dopamine receptor agonist might increase cardiac output and decrease cardiac rate both by increasing stroke volume and decreasing afterload. Such an agent might be useful in treating angina and hypertension.

Another cardiac response to catecholamine release is increased vulnerability to ventricular fibrillation. Recent studies (13) have shown that bromocriptine produced an increase of 50% in the ventricular fibrillation threshold in anesthetized dogs, and that pretreatment with the peripheral D-2 dopamine antagonist domperidone abolished this effect. This suggests that adrenergic induced cardiac arrhythmia may be inhibited by peripheral presynaptic dopamine agonists.

Benzazepine Chemistry

In our first attempts to identify potentially useful dopamine agonists, a number of compounds structurally related to dopamine were investigated. SK&F 38393 (I) (2,3,4,5-tetrahydro7,8-dihydroxy-1-phenyl-1H-3-benzazepine) was identified as an interesting lead (14,15,16) in that it was a dopaminergic selective renal vasodilator with no significant effects on blood pressure or heart rate at effective renal vasodilator doses. It also had significant diuretic and natriuretic effects. It was also a potent central dopamine agonist, but did not cause stereotypy, emesis, inhibition of prolactin release, or changes in dopamine turnover. This unique profile led us to study this class of compounds in some detail, and in this paper we wish to compare the properties of SK&F 38393 to those of several structurally closely related, but pharmacologically distinct, benzazepines (II,III,IV) and to dopamine.

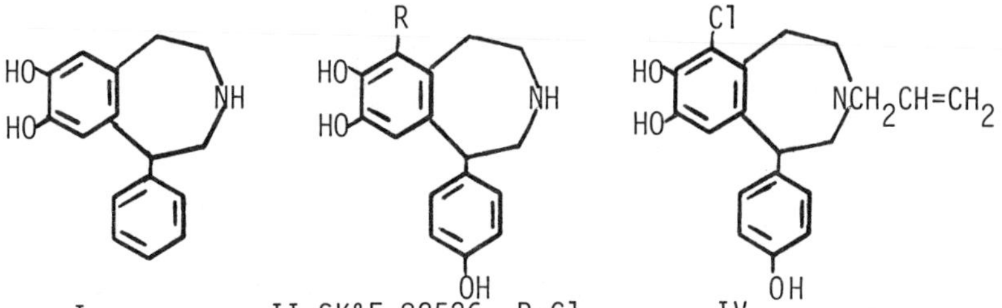

I	II,SK&F 82526, R=Cl	IV
SK&F 38393	III,SK&F 87516, R=F	SK&F 85174

The synthesis of I and II have been described (<u>17</u>). The fluoro analog III was prepared from 2-fluorohomoveratrylamine (<u>18</u>) as shown in Scheme I.

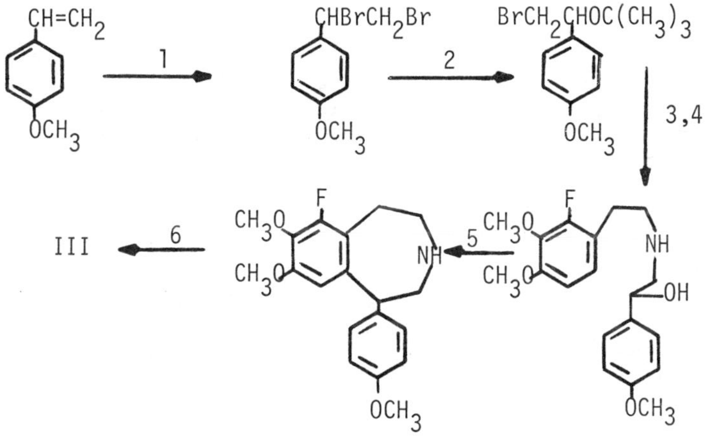

1. Br$_2$, dioxane, CH$_2$Cl$_2$; 2. <u>t</u>-BuOH, K$_2$CO$_3$,MgSO$_4$; 3. 2-fluoro-3,4-dimethoxyphenethylamine, DMF, K$_2$CO$_3$; 4. H$_2$SO$_4$, H$_2$O; 5. CH$_3$SO$_3$H, TFA; 6. BBr$_3$, CH$_2$Cl$_2$

Scheme I

The N-allyl compound IV was prepared by the sequence shown in Scheme II.

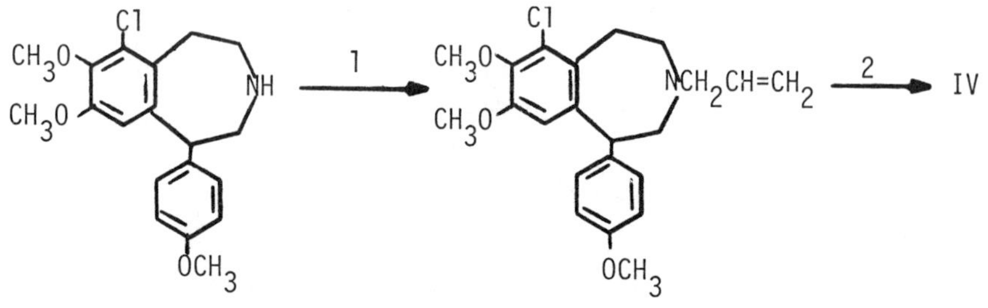

CH_3O —[benzazepine structure with Cl, OCH_3, OCH_3, NH, and p-methoxyphenyl]— $\xrightarrow{1}$ CH_3O —[benzazepine structure with Cl, OCH_3, OCH_3, $NCH_2CH=CH_2$, and p-methoxyphenyl]— $\xrightarrow{2}$ IV

1. $CH_2=CHCH_2Br$, K_2CO_3, DMF-acetone;   2. BBr_3, CH_2Cl_2

Scheme II

The solid state structure of the R-enantiomer (19,20) of II as its hydrobromide salt has been determined by the single crystal x-ray crystallographic technique. Several computer drawn projections derived from this data are shown in Figure 1. Among the interesting findings were the two almost parallel planes, one of which includes the benzo ring and the 1- and 5-carbons, and the other which includes the nitrogen and the 2- and 4-carbons.   The phenyl group is equatorial, and its plane is at a sharp angle to that of the benzo-fused ring. The carbon-oxygen bond length of the 8-hydroxy is significantly longer than those of the other two phenolic groups. This implies that the 7- and 4-phenyl carbon-oxygen bonds have more double bond character, and that the corresponding phenols are more acidic than the 8-phenol.

Benzazepine Pharmacology

A comparison of some of the pharmacological properties of the benzazepines with those of dopamine are shown in Table I. The methods used have been described in previous publications from our laboratories (14-17,21,22,23).
SK&F 38393 is about ten times less potent than dopamine as a renal vasodilator in anesthetized dogs, and shows only about half the maximal renal vasodilator effect. However, while at high doses dopamine increased mean arterial pressure and heart rate, high doses of SK&F 38393 increased only blood pressure. In contrast, SK&F 82526 is about ten times as potent as dopamine, and has about 1.6 times the maximal effect of dopamine as a renal vasodilator in anesthetized dogs. It causes a slight tachycardia, and at higher doses moderately decreases blood pressure. SK&F 87516 is about three times as potent as, and shows about 80% of the maximal effect of dopamine as a renal vasodilator. However, it produces negligible blood pressure effects and exhibits a potent and profound bradycardiac effect suggesting that it may have presynaptic dopamine agonist activ-

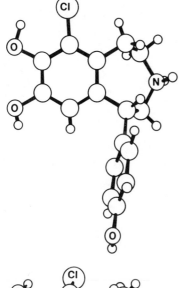

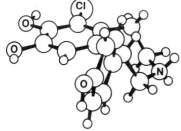

Figure 1. Computer-generated perspective drawings of the x-ray model of the hydrobromide salt of the R isomer of SK&F 82526.

Table I

Pharmacological Comparison of Dopamine and Benzazepines

Test[a]	Dopamine	SKF 38393 (I)	SKF 82526 (II)	SKF 87516 (III)	SKF 85174 (IV)
Selective Renal					
Vasodilator – Dog					
ED_{15}, decrease RVR, μg/kg (n)	2.7(3)	31(3)	0.56(4)	0.83(4)	7.7(3)
Avg. Max. decrease RVR, %	36	19	50	28	29
IVR (ED_{30})/RVR (ED_{15})	56	15	>10700	>7306	>780
MABP (ED_{20})/RVR (ED_{15})	113	10	-3300	>7306	>780
HR (ED_{20})/RVR (ED_{15})	141		>10700	-7	>780
Contralateral Rotation –					
Lesioned Rat					
RD_{500} mg/kg i.p. (i.p. dose and no. of turns)	—	0.7	(10 mg, 14 ± 7)	—	(2 mg, 297 ± 64)
RD_{500} μg/rat, i.c.	0.10	0.18	0.5	—	(10 mg, 647 ± 84)

Rat Striatal Adenylate Cyclase

EC$_{50}$, M	3.5×10^{-7}	7.1×10^{-8}	1.8×10^{-8}	2×10^{-7}	1.5×10^{-8}
% of 50 μM DA response (M)	100	68 (10^{-5})	80 (10^{-6})	75 (10^{-5})	85 (3.3×10^{-7})

Rat Caudate Specific Spiroperidol Binding

IC$_{50}$, M	5.34×10^{-6}	$3.44 \ 10^{-5}$	1.03×10^{-6}	4.56×10^{-6}	1.52×10^{-7}

Spontaneously Hypertensive Rat

Dose — mg/kg	0.5 i.p.	25 i.p.	50 i.p.	50 i.p.	25 i.p.	50 i.p.	25 p.o.	50 p.o.
Natriuretic Activity (3 hr)[b]	I	A	A	I	A	A	I	I
Antihypertensive Effect[b]	I	A	—	A	I	I	I	I

a Abbreviations are: RVR, renal vascular resistance; n, number of dogs; IVR, iliac vascular resistance; MABP, mean arterial blood pressure; HR, heart rate; RD$_{500}$, dose causing 500 turns in 2 hrs. Results in this test are reported as I (inactive) or A (statistically significantly active)

ity. Although many N-substituted benzazepines are not renal vasodilators (24), the N-allyl derivatives are an exception. Thus, SK&F 85174 (IV) is about one-third as potent and has about 75% of the maximal effect of dopamine as a renal vasodilator in anesthetized dogs. This compound also caused a significant decrease in heart rate in the presence of a significant decrease in blood pressure, again suggesting presynaptic dopaminergic activity.

One of the desired properties of a renal vasodilator for use in hypertension is a lack of central nervous system effects. Contralateral rotation in rats with unilateral 6-hydroxydopamine lesions of the substantia nigra is a measure of central dopaminergic activity (15). Dopamine given intracaudally (i.c.) is very potent in causing this rotation. SK&F 38393 is active both i.p. and i.c. demonstrating that it may have dopamine agonist activity and that it crosses the blood brain barrier. However, SK&F 82526, although potent when given i.c., is not active when given i.p. suggesting that it does not cross the blood brain barrier. SK&F 85174 shows very moderate activity when given i.p. Apparently, the combination of both the hydroxyphenyl and the secondary amine is required to prevent these benzazepines from crossing the blood brain barrier.

An important criteria of D-1 dopamine agonist activity is stimulation of dopamine-sensitive adenylate cyclase (8). Table I compares the activity of the four benzazepines and dopamine as stimulants of rat striatal adenylate cyclase. All of the benzazepines are only partial agonists with SK&F 82526 and SK&F 85174 being about 20 times more potent than dopamine and SK&F 38393 and SK&F 87516 being similar in potency to dopamine.

The rat striatal adenylate cyclase stimulation effects do not parallel the renal vasodilator effects. In fact, the 6-iodo and 6-methyl analogs of SK&F 82526 have only slight adenylate cyclase stimulating activity (they show adenylate cyclase inhibitor effects), but are potent renal vasodilators: 6-iodo analog, ED_{15} = 4.2 µg/kg, max. decrease RVR = 41%; 6-methyl analog, ED_{15} = 2.8 µg/kg, max. decrease RVR = 41%. This does not prove that renal vasodilation is not adenylate cyclase coupled. It does suggest that dog renal and rat striatal adenylate cyclase stimulants may have different structural requirements, and emphasizes the importance of biochemically characterizing the renal dopamine receptor(s), even though this may be much more difficult than similar characterization of other dopamine receptors (7,25).

Also shown in Table I is diuretic and antihypertensive activity in a spontaneously hypertensive rat screen. Dopamine was not significantly active at 0.5 mg/kg i.p. presumably due to its short biological half-life. SK&F 38393 is active both as a natriuretic and an antihypertensive agent. The diuretic activity has been characterized in other studies (26,27). It is natriuretic given i.p. or orally to saline-loaded, water-loaded, and sodium deficient rats. When SK&F 38393 was given orally at 20 mg/kg to conscious mannitol-phosphate infused dogs in a renal clearance procedure (23) sodium excretion increased 39%, renal plasma flow as measured by the clearance of para-aminohippurate (PAH) increased 46%, while glomerular filtration rate as measured by creatinine clearance did not change significantly.

Possibly relevant to the natriuretic activity of SK&F 38393 is emerging evidence that dopamine may be a physiological regulator of aldosterone synthesis (28). Dopamine receptors in a particulate fraction of calf adrenal glomerulosa cells have been shown to bind ^{3}H-2-amino-1,2,3,4-tetrahydro-6,7-dihydroxynapthalene (^{3}H-ADTN) in a specific (displaceable by 10 μM dopamine) saturable manner (29) (K_d, 0.29 mM; maximal binding capacity, 39 fmol/mg protein). SK&F 38393 is 50 times more potent than bromocriptine in competing for ADTN binding. In isolated collagenase dispersed calf adrenal glomerulosa cells, the K_d for dopamine, SK&F 38393, and bromocriptine were 5.1, 8.7, and 511 x 10^{-6}M respectively, and they decreased aldosterone release by 35, 42, and 2% (29).

In the spontaneously hypertensive rat (SHR) SK&F 82526 was inactive as a natriuretic, but was active as an antihypertensive agent. The weaker natriuretic activity of SK&F 82526 compared to SK&F 38393 was also seen in conscious dog renal clearance studies (27). SK&F 82526 given 5 mg/kg p.o. increased renal plasma flow 81%, glomerular filtration rate 15%, but elevated sodium excretion only about half the amount observed with SK&F 38393, and this in a somewhat delayed response. SK&F 82526 when given in the drinking water at 200 mg/kg/day for 30 days in developing SHR blunted the onset of hypertension and also lowered blood pressure acutely in anesthetized SHR. It did not lower blood pressure in anesthetized Dahl salt sensitive hypertensive rats.

In contrast to SK&F 82526, SK&F 87516 showed good natriuretic effects, but no significant antihypertensive effects in the conscious SHR after acute administration. SK&F 85174 showed neither natriuretic nor antihypertensive activity after oral or i.p. dosing under these conditions.

A very intriguing observation noted above was the possibility of presynaptic dopamine agonist activity with SK&F 87516 and SK&F 85174. The inhibitions of the constrictor response and the ^{3}H-norepinephrine release induced by brief intermittent periarterial sympathetic nerve stimulation of the perfused rabbit ear artery is a convenient in vitro assay for presynaptic dopamine agonist activity (30). In this preparation, both SK&F 85174 and SK&F 87516 inhibited both stimulation responses with an EC_{50} of about 100 nM. Another measure of D-2 activity is inhibition of specific spiroperidol binding (20). Comparison of these data (Table I) showed that SK&F 85174 is thirty times as potent as dopamine, and thus is a potent D-2 agonist. However, the activity in adenylate cyclase stimulation indicates that this compound may be best described as a D-1/D-2 agonist (19, 20).

SK&F 85174 was chosen for more detailed study (31). In anesthetized dogs, tachycardia induced by cardiac accelerator nerve stimulation was inhibited by intravenous SK&F 85174 (ED_{50}, 42 μg/kg) which effect was antagonized by metoclopramide and 1-sulpiride. In anesthetized rats (normotensive, spontaneously hypertensive, DOCA-salt hypertensive) SK&F 85174 decreased both blood pressure and heart rate. The results suggested that SK&F 85174 has presynaptic dopaminergic activity and may be useful in the treatment of angina and hypertension (31).

The interesting renal vasodilator effect of SK&F 82526 has led to a study of its renal effects in normal volunteers under conditions of water diuresis (32). The compound was studied at single oral doses of 25, 50 and 100 mg. Dose related changes were observed in renal plasma flow as measured by PAH clearance: at 100 mg this increased 51% at two hours, the time of peak drug effect. Inulin clearance was unchanged demonstrating that, as in the dog, the compound did not alter glomerular filtration rate. However, at 100 mg sodium and water excretion were increased 89 and 54%, respectively, above control. Surprisingly, potassium excretion did not increase significantly. These results suggested that SK&F 82526 might have clinical utility in the prophylaxis and treatment of acute renal failure.

Discussion

This brief overview of the pharmacological activity of four closely related benzazepines has shown that relatively small structural changes have given quite distinct pharmacological profiles. Thus starting from SK&F 38393, introduction of a 6-chloro substituent increases potency as a renal vasodilator by a factor of about ten, and potency in stimulating adenylate cyclase by a factor of about eight (17). Finally, addition of the 4-hydroxyl to the phenyl increases potency as a renal vasodilator by another factor of ten, but does not cause a change in adenylate cyclase stimulation. This may be rationalized by assuming that the 6-chlorine substituent increases the acidity of the 7-hydroxyl significantly, possibly not only increasing binding to the receptor, but also orienting it in a more specific manner. The addition of the 4-hydroxyl on the phenyl may also cause additional binding at least to a renal receptor, but apparently not to the rat striatal adenylate cyclase receptor.

Fluorine substitution in place of chlorine (compare SK&F 87516 to SK&F 82526) seems even more trivial than the changes discussed above, but it apparently enhances diuretic activity and introduces substantial peripheral presynaptic (D-2) activity. Since the renal vasodilator potency is essentially unchanged, this suggests that diuretic activity is determined by receptors with different properties than the renal vasodilator

receptors. An important difference between fluorine and chlorine is size. Perhaps in the D-2 and dopamine diuretic receptors
the space for a properly oriented benzazepine agonist is so restricted that fluorine fits, but chlorine does not. Another
difference between the fluoro and chloro groups is their effect
on the acidities of the catechol hydroxyls. Based on the reported ionization constants of the corresponding 2- and 3-halo-
phenols in water (2-F, pKa 8.81; 2-Cl, 8.41; 3-F, 9.28; 3-Cl,
9.02)(33, 34), it is probable that the chlorocatechol is more
acidic than the fluorocatechol. In addition, it has been shown
(35) that intramolecular hydrogen bonding occurs in 2-chloro-
phenol, but not in 2-fluorophenol.

The N-allyl group of SK&F 85174 is another group with a
large influence on dopaminergic profile. It allows renal vasodilator activity and stimulation of adenylate cyclase, while
most N-alkyl groups cause inhibition of renal vasodilator activity and adenylate cyclase (24). It induces potent D-2 activity in contrast to the selective D-1 profile of SK&F 82526.
A possible explanation is that in the presence of the allyl
group, the nitrogen electron pair is delocalized due to homo-
allylic interaction of the n, π type (36)

thus allowing binding without precise apposition of the nitrogen and the receptor base binding site. The N-allyl group reduces renal vasodilator potency 25-fold when it replaces the NH
of SK&F 82526, but this may also be caused in part by the more
lipophilic nature of the compound causing more widespread distribution in the body, thus decreasing renal concentration.

In summary, each of these four structurally closely related
benzazepines has a unique combination of dopaminergic activities. This suggests that the various receptors involved do not
differ too greatly from each other, as indeed could have been
surmised from the fact that all are activated by dopamine. However, they do differ sufficiently so that selectivity between
them is possible. It is possible that some of these compounds
may prove to be therapeutically useful, and may also be useful
as tools for greater understanding of dopamine pharmacology.

Literature Cited

1. Guyton, A.C.; Colman, T.G.; Cowley, A.W.; Manning, R.D.;
    Norman R.A.; Ferguson, J.D. *Circ. Res.* 1974, 35, 159.
2. Goldberg, L.I. *Pharmacol. Rev.* 1972, 24, 1.

3. Setler, P.E.; Pendleton, R.G.; Finlay, E. J. Pharmacol.
    Exptl. Therap. 1975, 192, 702.
4. Pendleton, R.G.; Finlay, E.; Sherman, S. Naunyn-Schmiede-
    bergs Arch. Pharmacol. 1975, 289, 171.
5. Hahn, R.A.: Wardell, J.R. Naunyn-Schmiedebergs Arch.
    Pharmacol. 1980, 314, 177.
6. Goldberg, L.I.; Kohli, J.D. Comm. in Psychopharmacol.
    1979, 3, 447.
7. Kotake, C.; Hoffman, P.C.; Goldberg, L.I.; Cannon, J.G.
    Mol. Pharmacol. 1981, 20, 429.
8. Kebabian, J.W.; Calne, D.B. Nature 1979, 277, 93.
9. Goldberg, L.I.; Weder, A.B. "Recent Advances in Clinical
    Pharmacology"; Turner, P.; Shand, D.G. ed.; McMillan, New
    York, 1980, Vol. 2, 149.
10. Langer, S.Z. Pharmacol. Rev. 1981, 32, 337.
11. Lokhandwala, M.F.; Jandhyala, B.S. J. Pharmacol. Exptl.
    Therap. 1979, 210, 120.
12. Clark, B.J. Postgraduate Med. J. 1981, 57 (Suppl. 1), 45.
13. Falk, R.H.; Desilva, R.D.; Lown, B. Cardiovas. Res. 1981,
    15, 175.
14. Pendleton, R.G.; Sander, L.; Kaiser, C.; Ridley, P.T.
    European J. Pharmacol. 1978, 51, 1.
15. Setler, P.E.; Sarau, H.M.; Zirkle, C.L.; Saunders, H.L.,
    European J. Pharmacol. 1978, 50, 419.
16. Wilson, J.W. Abstracts, 16th National Medicinal Chemistry
    Symposium, American Chemical Society 1978, p. 155.
17. Weinstock, J.; Wilson, J.W.; Ladd, D.L.; Brush, C.K.;
    Pfeiffer, F.R.; Kuo, G.Y.; Holden, K.G.; Yim, N.C.F.;
    Hahn, R.A.; Wardell, J.R.; Tobia, A.J.; Setler, P.E.;
    Sarau, H.M.; Ridley, P.T. J. Med. Chem. 1980, 23, 973.
18. Ladd, D.L.; Weinstock, J. J. Org. Chem. 1981, 46, 203.
19. Kaiser, C.; Dandridge, P.A.; Weinstock, J.; Ackerman, D.M.;
    Sarau, H.M.; Setler, P.E.; Webb, R.L.; Horodniak, J.W.,
    Matz, E.D. Acta Pharma. Seuc. in press.
20. Kaiser, C.; Dandridge, P.A.; Garvey, E.; Hahn, R.A.; Sarau,
    H.M.; Setler, P.E.; Bass, L.S., Clardy, J. J. Med. Chem.
    1982, 25, 697.
21. Hahn, R.A.; Wardell, J.R. J. Cardiovasc. Pharmacol. 1980,
    2, 583.
22. Hahn, R.A.; Wardell, J.R.; Sarau, H.M.; Ridley, P.T. J.
    Pharmacol. Exp. Ther. 1982, 223, 305.
23. Ackerman, D.M.: Weinstock, J.; Wiebelhaus, V.D.; Berkowitz,
    B. Drug Dev. Res. 1982, 2, 283.
24. Pfeiffer, F.R.; Wilson, J.W.; Weinstock, J.; Kuo, G.Y.;
    Chambers, P.A.; Holden, K.G.; Hahn, R.A.; Wardell, J.R.;
    Tobia, A.J.; Setler, P.E.; Sarau, H.M. J. Med. Chem. 1982,
    25, 352.
25. Felder, R.; Pelayo, J.; Belcher, M.; Calcagno, P.; Eisner,
    G.; Jose, P. Clin. Res. 1981, 29, 462A.

26. Brennan, F.T.; Sosnowski, G.; Erickson, R.; Mann, W.; Sulat, L.; Wiebelhaus, V.D. Fed. Proc. 1979, 38, 748.
27. Mann, W.A.; Sosnowski, G.F.; Kavanagh, B.J.; Erickson, R.W.; Brennan, F.T.; Wielbelhaus, V.D. Fed. Proc. 1981, 40, 647.
28. Sowers, J.R.; Berg, G.; Martin, V.S.; Moyes, D.M. Endocrinol. 1982, 110, 1173.
29. Bevilacqua, M.; Vago, T.; Malago, E.; Norbiato, G. Abstracts, Symposium on Dopamine Receptor Agonists, Stockholm 1982.
30. Steinsland, O.S.; Hieble, J.P. Science 1978, 199, 443.
31. Blumberg, A.L.; Hieble, J.P.; McCafferty, J.; Hahn, R.A.; Smith, J. Fed. Proc. 1982, 41, 1345.
32. Stote, R.M.; Erb, B.; Alexander, F.; Givens, K.; Familiar, R.; Dubb, Jr. Kidney Int. 1982, 21, 248.
33. Bennett, G.M.; Brooks, G.L.; Glasstone, S. J. Chem. Soc., 1935, 1821.
34. Judson, C.M.; Kilpatrick, M. J. Am. Chem. Soc., 1949, 71, 3110.
35. Allen, E.A.; Reeves, L.W. J. Phy. Chem. 1962, 66, 613.
36. Morishima, I.; Yeshikawa, K.; Hashimoto, M.; Bekki, K. J. Am. Chem. Soc. 1975, 97, 4283.

RECEIVED February 16, 1983

Commentary: Dilemmas in the Synthesis of Clinically Useful Dopamine Agonists

JAMES Z. GINOS

Memorial Sloan-Kettering Cancer Center, Cotzias Laboratory of
Neuro-Oncology, New York, NY 10021

The authors have described four structurally related dopaminergic compounds with pharmacological profiles sufficiently
different from each other to permit them to claim that at least
one of them has high stereoselectivity for D-2 dopamine (DA) receptors as defined by Kebabian and Calne (1). All four of these
compounds have been identified as partial DA agonists, a feature
they characteristically share with apomorphine (2), now a classical standard DA agonist often used for purposes of comparison
with newly synthesized DA agonists. All four of these compounds
have been shown to be peripherally active on DA receptors, as
evidenced by their cardiovascular and renal vascular effects
(lowering of renal vascular resistance and increasing renal blood
flow). With the exception of the untested fluoro derivative, of
the remaining compounds, not surprisingly, three have been shown
to also have central DA effects, although one of them had to be
administered intracerebrally to demonstrate such effects, since,
according to the authors, in contrast to the other two compounds,
it fails to cross the blood-brain barrier (BBB).

The authors start out by listing in a logical fashion the
desired characteristics of a DA agonist if this is to be clinically applicable as an antihypertensive drug. It should be an
orally active drug, free of nausea-causing effects, a property
with which DA itself and many other DA agonists are associated,
and, above all, should be free of undesirable pressor and cardiac
effects associated with stimulants of peripheral β- and α-adrenergic receptors. As the authors have pointed out, DA itself,
although a powerful DA agonist, is capable of acting on β- as
well as on α-receptors.

Thus, the aim of the present-day investigator who strives to
conceive and synthesize new, more effective DA-like drugs against
hypertension, is to dissect out those molecular features of the
drug, or pharmacophores, or perhaps at least modify them, in an
effort to achieve high specificity; that is a pure DA agonist,
free of adrenergic stimulating properties. Although he may not
be able to achieve this in a rational manner, he may still

0097–6156/83/0224–0170$06.00/0

succeed in attenuating such undesirable properties to the point
where these become less worrisome to the clinician, at least in
the dose range where these DA agonists can still function as
therapeutic agents.

However, what complicates the problem for the investigator
is the fact that some of the properties related directly to DA
agonism may be from the clinican's as well as the patient's point
of view equally undesirable. I particularly refer to the emetic
effects which often accompany the administration of such drugs.
Some of the alkaloids, α-bromocriptine (3), lergotrile (4), apomorphine and its congener N-(n-propyl)-norapomorphine (5), as
well as L-dopa (6)-- the biochemical precursor of DA itself-- all
of which have been used clinically in the treatment of disorders
shown to be amenable to dopaminergic action, have all caused
nausea in patients to a varying degree, particularly in parkinso-
nians. This untoward effect can be traced to the stimulation of
DA receptors in the area postrema center (7, 8). More recently,
N-(n-propyl)-N-(n-butyl)-DA, an N,N-dialkylated DA derivative,
shown to act on D-2 as well as D-1 receptors, was also shown to
be free of β-adrenergic properties and to possess only weak
α-stimulating attributes at the upper dose ranges (9). Nevertheless, it typically caused in a dose-dependent manner nausea in
dogs as well as in humans, both normal and hypertensive (10).

Moreover, the majority of the drugs with DA-like action
tested in intact animals and capable of crossing the BBB, have
shown centrally-mediated behavioral effects, such as stereotypy
(gnawing, sniffing, rearing, pecking in doves) and elevated
locomotor activity (11). As a matter of fact, these behavioral
effects have been used as a basis for recognizing and even
quantitating, albeit crudely, dopaminergic action. However, to
the best of this author's knowledge, no one has been able to
correlate convincingly such animal behavioral effects with centrally-mediated behavioral effects in human subjects undergoing
long-term treatment with DA agonists, though, as it is well
known, parkinsonian patients have shown, often in a dose-dependent manner, hallucinatory and other mental aberrations (12) and,
depending on the duration of treatment, involuntary movements
(buccolingual and/or jerk-like movements of the limbs).

Thus, the clinician who wishes to use an orally active,
antihypertensive drug that mediates its therapeutic effects via
dopaminergic mechanisms, even if the medicinal chemist has been
able to eliminate or modify those molecular structural features
responsible for the adrenergic stimulatory effects, has to cope
with the aforementioned undesirable effects. In principle, there
are a number of approaches the clinician may use in dealing with
this problem. He may devise a drug regimen in which the administration of the drug may be carried out in gradually increased
doses until the dose attained achieves its desirable effect
before untoward effects emerge or, if they do, their intensity
is minimal. Or, he may resort to the coadministration of a DA

antagonist, assuming that such does exist, tailored pharmacologically to antagonize the unwanted effects. This, of course, may
imply that there are specific DA receptors responsible for such
effects. One, naturally, cannot exclude the speculation that
such untoward effects, particularly those of mental aberration
and involuntary movements are the consequences of upsetting the
delicate balance of interdependent neurotransmission mechanisms
by overloading the cerebral dopaminergic circuitry at the expense
of other circuitries, such as the cholinergic one, for example.
If this is true, then in principle one might be able to reduce
the imbalance by administering to the patient the appropriate
non-dopaminergic neurotransmitter or one that mimics it or arti-
ficially induces its release in an effort to redress such an
imbalance. The problem becomes more complicated with such an
approach and is frequently fraught with difficulties and
potential problems.

The other remaining approach is that of the medicinal
chemist who may attempt to modify the effective DA agonist in a
manner that it becomes impermeable to the BBB without affecting
its therapeutic attributes. Thus, in one stroke he may eliminate
all the possible centrally mediated effects. However, there
still remains the problem of emesis and nausea, since the area
postrema center remains outside the BBB and therefore is accessible to any peripherally active DA agonist. There are two
possible solutions to this problem. One may search for a peripherally active DA antagonist which blocks specifically the DA
receptor in the area postrema center. That, of course, implies
that the postrema DA receptor is intrinsically different from
other peripheral DA receptors, at least to the extent that one
may pharmacologically differentiate it from the other ones. The
proliferation of classes and subclasses of DA receptors is now
part of the present state of art and one's belief regarding the
proper classification may depend on whether one chooses strictly
pharmacological, biochemical, receptor-binding or behavioral
criteria, or any combination thereof. This author, for heuristic
considerations and reflecting his own predilection, perhaps,
rather than for strictly scientific reasons, will assume that
there are two broad classes of DA receptors: D-1 and D-2. If
one identifies the area postrema DA receptor with that of the D-2
type (11), then it becomes apparent that the use of an antagonist
to block the area postrema DA receptor will inevitably eliminate,
or at least reduce substantially any pharmacological benefits
derived from the stimulation of peripheral D-2-type receptors.
Thus, the medicinal chemist and pharmacologist are faced with an
intractable problem, unless the medicinal chemist can create a
drug with unique physical attributes that give it preferential
access to peripheral vascular D-2 receptors or to the area
postrema receptor if intended as a blocker. The problem is a
formidable one, but then there is always the ingenious
alternative of compromise, balancing the adverse effects against
the expected benefits.

Literature Cited

1. Kebabian, J. W.; Calne, D. B. Nature 1979, 277, 93-96.
2. Goldberg, L. I. "Advances in Biochemical Psychopharmacology",
   Vol. 19; Roberts, P. J.; Woodruff, G. N.; Iversen, L. L.,
   Eds.; Raven Press; New York, 1978, pp 119-129.
3. Calne, D. B.; Teychenne, P. F.; Claveria, L. E.; Eastman, R.;
   Greenacre, J. K.; Petrie, A. British Med. J. 1974, 23,
   442-444.
4. Lieberman, A.; Miyamoto, T.; Battista, A. F.; Goldstein, M.
   Neurology 1975, 25, 459-462.
5. Cotzias, G. C.; Papavasiliou, P. S.; Fehling, C.; Kaufman,
   B.; Mena, I. N. Engl. J. Med. 1970, 282, 31-33.
6. Cotzias, G. C.; Papavasiliou, P. S.; Gellene, R. N. Engl. J.
   Med. 1969, 280, 337-345.
7. Borison, H. L.; Rosenstein, R.; Clark, W. G. J. Pharmacol.
   Exp. Ther. 1960, 130, 427-430.
8. Sourkes, T. L.; Samarthji, L. in Agranoff, B. W.; Aprison,
   M. H., Eds.; "Advances in Neurochemistry", Plenum Press, New
   York, 1975, 1, pp 247-299.
9. Kohli, J. D.; Weder, A. B.; Goldberg, L. I.; Ginos, J. Z.
   J. Pharmacol. Exp. Ther. 1980, 213, 370-374.
10. Taylor, A. A.; Young, J. B.; Brandon, T. A.; Ginos, J. Z.;
    Goldberg, L. I.; Mitchell, J. R. Circulation (in press).
11. Seeman, P. Pharmacol. Rev. 1980, 32, 230-313.
12. Cotzias, G. C.; Papavasiliou, P. S.; Ginos, J. Z.; Tolosa,
    E. S. "The Clinical Neurosciences"; Tower, D. B. (Ed.);
    Raven Press; New York, 1975, pp 323-329.

RECEIVED February 3, 1983

Dopamine Agonists and Antagonists in Duodenal Ulcer Disease

S. SZABO
Brigham and Women's Hospital and Harvard Medical School,
Departments of Pathology, Boston, MA 02115

J. L. NEUMEYER
Northeastern University, College of Pharmacy and Allied Health Professions,
Section of Medicinal Chemistry, Boston, MA 02115

Recent data in our laboratories indicated the
possible involvement of central and peripheral
dopamine binding sites in the pathogenesis of
duodenal ulceration. Structure activity studies
with duodenal ulcerogens implicated dopamine as a
putative mediator and/or modulator in duodenal
ulceration. Using the cysteamine- or propionitrile-
induced duodenal ulcer model in rats, we found that
dopamine agonists (e.g., bromocriptine, lergotrile,
L-DOPA, apomorphine and its derivatives) prevented,
while the antagonists (e.g., haloperidol, pimozide)
aggravated the experimental duodenal ulcers. These
drugs also modulated the output of gastric acid and
pepsin, as well as that of pancreatic bicarbonate
and enzymes. The administration of duodenal ulcerogens to rats changed the binding characteristics of
[^{3}H]-dopamine in the gastric and duodenal mucosa and
muscularis propria, as well as in brain regions.
Thus, certain dopamine agonists especially dopaminergic aporphine alkaloids may represent new
agents to prevent or treat duodenal ulcer disease.

The neurotransmitter dopamine has currently found clinical
application in the treatment of cardiovascular collapse and
shock. Treatment with dopamine-related drugs has been
limited to such brain disorders as Parkinson's disease,
schizophrenia, and hyperprolactinemias. Accumulating data from
animal experiments, however, indicate the possible involvement of
dopamine in other diseases and the potential use of dopamine
agonists or antagonists in these disorders. Gastrointestinal
disturbances (especially duodenal ulcer disease) seem to
represent such a group of dopamine-sensitive alterations.
    Duodenal ulcer is about four times more prevalent than
gastric ulcer in this part of the world (1,2). Until recently,
however, animal models have been available mostly for gastric

erosions and ulcers since almost any chemical in toxic doses
produces non-specific, stress ulcers (3,4). On the other hand,
easily and rapidly reproducible duodenal ulcers have not been
available until the introduction of propionitrile (5) and
cysteamine (6) models. These chemically induced duodenal ulcers
in the rat closely resemble the human duodenal ucler both
functionally and morphologically (7-10). Furthermore they serve
as good animal models to study the pathogenesis of duodenal
ulceration and to test new anti-ulcerogenic drugs.

The pathogenesis of duodenal ulcer disease is poorly
understood and its therapy is mostly empirical (1,2). The
therapy for ulcer disease has virtually been limited to
neutralization of gastric acid by antacids and inhibition of acid
secretion by anticholinergics and histamine H_2 receptor
antagonists (1,11). Yet, there is no inherent disorder or
imbalance of histamine or acetylcholine in any form of duodenal
ulcer disease. Besides the questionable "vagal hypertone" only
hypergastrinemia (e.g., in Zollinger-Ellison syndrome) in the
minority of peptic ulcer has been documented (1,2). It is
generally accepted that duodenal ulcer disease is a pluricausal
or multifactoral disorder where gastric acid is only one of the
many things to have gone astray. How can we then expect to have
the best, etiologic treatment for ulcer disease when only
antacids or antisecretory agents are used? Although the presence
of (increased?) acid in the stomach and duodenum is probably
necessary for the development of ulcers, the pathogenesis of
ulcer disease is much more complex than a simple chemical
neutralization or depressed secretion of acid.

In this review we shall present data from our and other
laboratories and argue that dopamine has as yet an unrecognized
etiologic role in the pathogenesis of duodenal ulcer disease.
Dopamine agonists may have multiple sites of action in this
disease (e.g., by affecting not only gastric acid and pepsin, but
also the duodenum itself, the pancreas, adrenals, and brain).
Nevertheless, the role of dopamine agonists in the treatment of
duodenal ulcer disease has not been explored. This review
presents structure-activity studies with chemical duodenal
ulcerogens, pharmacologic modulation with dopamine-related drugs
and biochemical studies with experimental duodenal ulcers. A
brief discussion and overview of implications of these findings
will conclude this chapter.

Structure-activity Relationship of Duodenal Ulcerogens

Experimental duodenal ulcers until recently, were not easily
induced in the most commonly used laboratory animals (e.g., rats,
mice)(12). The few available methods were complex and not widely
used. For example, the chronic deficiency of pantothenic acid in
certain strains of rats (13), 24-h sc infusion of secretagogues
(e.g., histamine, carbachol, pentagastrin) in fasted rats (14),

or local application of acetic acid on the duodenum (15) were found to cause duodenal ulcer but the incidence of these lesions was low and variable.

In 1972 we found that injections of propionitrile (Figure 1) consistently produced solitary, often perforating duodenal ulcers in the rat (5). These lesions occurred 3-5 mm from the pylorus of the stomach, mostly - as in humans - on the anterior wall of the duodenum. The ulcers developed 24-48 h after the initial administration of propionitrile and frequently penetrated into the liver or pancreas (5).

Cysteamine (Figure 1) was also found to produce duodenal ulcers, as well as adrenocortical necrosis in rats (6,16,17). The lesion in the duodenum developed even more rapidly (e.g., perforation in 24 h after a single dose) and more predictably than with propionitrile. Acetanilide was the first aryl chemical noted to cause duodenal ulcer (18). Subsequently, 3,4-toluenediamine (19) and 3,4-toluenedithiol (20) were also shown to induce duodenal ulcers and occasionally adrenal necrosis in rats.

The original findings and reports on the duodenal ulcerogenic action of chemicals such as propionitrile and cysteamine were independent observations. At the time no common structural, chemical, or biologic property of these chemicals was apparent. Subsequently, however, we recognized that the duodenal ulcerogenic property of these chemicals may be associated with a two-carbon (-C-C- or 2C) group bearing methyl, cyano, sulfhydryl or amino group(s) on one or both ends of the carbon chain (4,20-22). The length of the carbon backbone is an important feature for the duodenal ulcerogenic properties of such agents. Thus, we hypothesized that propionitrile (Figure 1) might be anomalous and that by prolonging the carbon chain with one carbon (i.e., obtaining n-butyronitrile) or by replacing the amino groups in 3,4-toluenediamine by thiols, we should have chemicals with possibly higher duodenal ulcerogenic potency than the original compounds. Both of these predictions proved correct (20).

Structure-activity studies also revealed that unsaturated derivatives (e.g., -C=C or -C-C=C, such as acrylonitrile, and various allyl compounds), without exception exerted higher adrenocorticolytic than ulcerogenic actions (20-22). Addition of a new methylene group to the carbon chain of propionitrile (CH_3CH_2CN), resulting in n-butyronitrile ($CH_3CH_2CH_2CN$), enhanced the ulcerogenic effect, whereas addition of a bromine or nitrile group at this site decreased, and an amino group abolished, the ulcerogenic action. Thus, the ulcerogenic potency of these radicals may be presented as follows: $-CH_3 > -CN > -Br > -Cl > -NH_2 -OH$ (20,21). The summary of our work on structure-activity studies with 56 alkyl chemicals is presented in Figure 2. Compilations of data with terminal nitriles, amines and thiols revealed that chains with two or three carbons had the highest duodenal

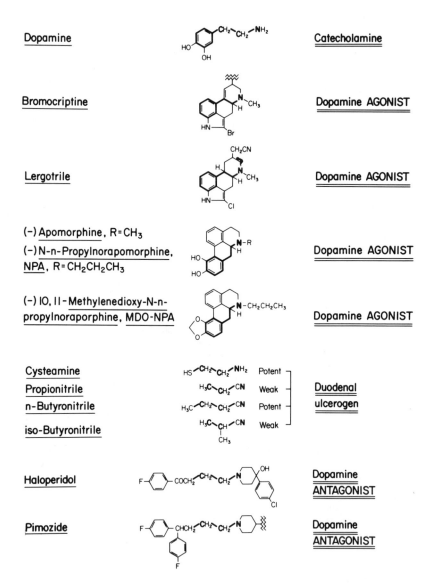

*Figure 1.   Structure of dopamine, dopamine-related drugs, and chemical duodenal
ulcerogens.*

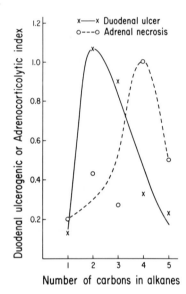

Figure 2. Duodenal ulcerogenic and adrenocorticolytic activities of terminal nitriles, amino and thiol alkanes. (Reproduced with permission from Ref. 34. Copyright 1980, International Institute of Stress.)

ulcerogenic index (Figure 2). The peak of adrenocorticolytic
index suggested a slight (probably one carbon) "shift" to the
right (Figure 2).

Among aryl chemicals tested, the duodenal ulcerogenic
activity of 3,4-toluenedithiol was predicted on the basis of our
initial structure-activity analysis of the ulcerogenic properties
of propionitrile, cysteamine, and 3,4-toluenediamine (20,21).
Indeed, replacement of amines by thiols in 3,4-toluenediamine not
only preserved the ulcerogenic property, but also slightly
enhanced potency. On the other hand, enlargement of the space
between the substituents in positions 3 and 4 in toluene or
benzoic acid (e.g., to 2,4, or 2,5, or 2,6, or 3,5) abolished the
duodenal ulcerogenic activity. Toluene alone is not an
ulcerogen. However, the addition of an amino group in a certain
position (e.g., o-toluidine and p-toluidine) results in prominent
duodenal ulcerogenic activity (4). Nevertheless, besides the
quality and position of substituents, other factors probably also
have a role in the causation of experimental duodenal ulcers,
e.g., the vicinal amino groups in 1,2-diaminocyclohexane,
pyridine, or pyrimidine failed to be associated with ulcerogenic
properties. Surprisingly, 4,5-diaminopyrimidine or
1,3-dicyanobenzene exerted mild to moderate duodenal ulcerogenic
actions (4).

An interesting observation in our structure-activity studies
was that ethylamine, a part of the histamine molecule,
4-(2-ethylamine)imidazole, known for a long time to be a potent
stimulator of gastric secretion, also produced duodenal ulcers in
rats (20,22). Furthermore, structural similarities also exist
between dopamine and histamine H_2 receptor antagonists metiamide
and cimetidine (23,24).

The duodenal ulcerogenic action of propionitrile and
cysteamine was confirmed and extended by several authors (7,8,
25-32). The model, especially the rapidly developing
cysteamine-induced duodenal ulcer, was actually introduced in
several laboratories to routinely test drugs for anti-ulcer
activity (7,8,27,33). The cysteamine model, a simple, fast, and
inexpensive procedure, became especially useful after a chronic
form of duodenal ulcer was developed (9) and similarities between
the model and human duodenal ulceration were noted (7-10).

The pathogenesis of these experimental duodenal ulcers is
poorly understood. The role of gastric acid, gastrin,
somatostatin and neurotransmitters, delayed gastric emptying and
the early and late morphologic changes in the duodenum are
extensively discussed in recent reviews and original papers
(4,22,31,34-37). The most relevant extrapolation from structure-
activity studies, however, is the possibility of identifying
putative endogenous modulators and/or mediators of duodenal ulcer
disease. On the basis of previous discussions we suggested that
dopamine might be one such neurotransmitter (36). Consequently,
because of similarity between dopamine, chemical duodenal

ulcerogens, dopamine agonists and antagonists, we hypothesized that dopamine agonists might prevent while antagonists aggravate the cysteamine- or propionitrile-induced duodenal ulcers (36) (Figure 1). The results of these and new pharmacologic studies are presented in the next section.

Pharmacologic Modulation with Dopamine-related Drugs

These studies were also performed on Sprague-Dawley female rats which had unlimited access to Purina laboratory chow and tap water. The details of these experiments have been described elsewhere (36). Pretreatment with propylene glycol (solvent for bromocriptine) had no effect on the development of acute duodenal ulcers produced by cysteamine, the dose of which was chosen to exert submaximal ulcerogenic effect in order to detect other preventing and aggravating actions among dopamine-related drugs (Figures 3,4). On the other hand, pretreatment with either bromocriptine or lergotrile showed a dose- and time-dependent anti-ulcerogenic action (Figures 3,4). In fact, 5 mg of bromocriptine or lergotrile given once daily for 4 days was significantly more effective than a single dose administered 30 min prior to cysteamine, and 21-day pretreatment was significantly better than the 4-day course. Mortality in the cysteamine model was abolished after prolonged treatment with these ergot alkaloids. On the other hand, pretreatment with either haloperidol or pimozide aggravated both the duodenal ulcer index and the mortality after cysteamine administration (Figure 4) (36). Also in the cysteamine duodenal ulcer model, apomorphine and L-DOPA exerted an anti-ulcerogenic action in a much narrower dose range than the ergots (36).

The actions of these drugs are not specific to cysteamine since the intensity of propionitrile-induced acute duodenal ulcers was also significantly diminished after pretreatment with bromocriptine or lergotrile (Table I).

To gain insight into the mechanisms of these anti-ulcerogenic actions of dopamine agonists, experiments were designed to measure the secretions of acid and pepsin from the stomach, bicarbonate from the pancreas and duodenum, as well as the activity of pancreatic enzymes trypsin and amylase in rats with chronic gastric fistula (38). In these secretory studies both cysteamine and propionitrile increased gastric acid output and decreased duodenal neutralization of acid (38,39). A single dose of lergotrile or bromocriptine significantly suppressed both the initial and the total acid output in rats injected with propionitrile (Table II). Surprisingly, pretreatment with ergots for one week (which was associated with better anti-ulcerogenic treatment than a single dose), slightly diminished the propionitrile-induced initial acid output, but it did not modify the total 7 h acid output (Table II). Thus, we seem to be dealing with prominent duodenal anti-ulcerogenic actions not

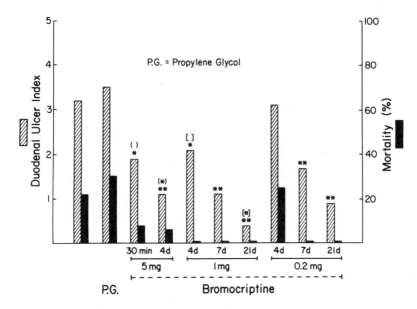

Figure 3.   Effect of bromocriptine on duodenal ulcer produced by cysteamine in the rat. (Reproduced with permission from Ref. 36. Copyright 1979, The Lancet Ltd.)

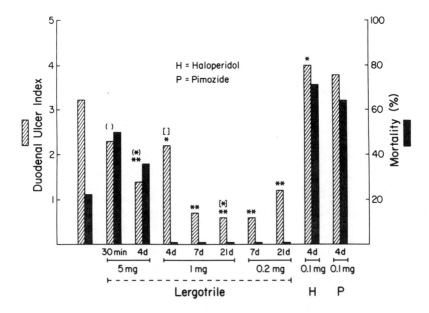

Figure 4.   Effects of lergotrile, haloperidol, or pimozide on duodenal ulcer produced by cysteamine in the rat. (Reproduced with permission from Ref. 36. Copyright 1979, The Lancet Ltd.)

TABLE I. Effect of Bromocriptine or Lergotrile on
Propionitrile-Induced Duodenal Ulcers in the Rat

| Group | Pretreatment # | Duodenal ulcer | | Mortality |
		Incidence (Pos./Tot.)	Intensity (Scale:0-3)	(%)
1.	None	16/17	1.2+0.2	88
2.	Bromocriptine	5/9*	0.6+0.3*	100
3.	Lergotrile	6/9	0.6+0.2*	75

\#   In addition, rats of all groups were given propionitrile 30
min after the administration of dopamine agonists.

* = p<0.5

(Reprinted from Ref. No. 59)

TABLE II. Effect of Lergotrile or Bromocriptine on Propionitrile-
Induced Gastric and Pancreatic/Duodenal Secretions in Conscious
Rats with Chronic Gastric Fistula

| Group | Pretreatment # | Acid output (μEq± SEM) | | Base output (μEq±SEM) |
		Initial 3hr	Total 7 hr	Init. 3h
1.	None	630+92	920+62	145+13
2.	Lergotrile (0.2mg, once)	104+57***	398+86**	74+24*
3.	Bromocriptine (0.2 mg, once)	69+32***	405+85**	148+46
4.	Lergotrile (0.2 mg, once daily for 1 week)	264+72*	814+129	142+19
5.	Bromocriptine (0.2 mg, daily for 1 week)	103+49***	818+221	135+30

\#In addition, rats of all groups were given propionitrile 30 min
after the administration of dopamine agonists.

* =p<0.05;   ** = p<0.01;   *** = p<0.001

(Reprinted from Ref. No. 59)

always accompanied by decreased gastric acid output. The base output was decreased only by lergotrile after the acute but not following chronic treatment (Table II). The output of pancreatic trypsin and amylase were suppressed by either bromocriptine or lergotrile.

A discrepancy between the anti-secretory and anti-ulcerogenic actions of bromocriptine and lergotrile was further indicated by a beneficial action of these ergots in chronic duodenal ulcer (Table III). On the first day of these experiments, rats were given an acute duodenal ulcerogenic regimen of cysteamine and on the second day the animals were placed on drinking water containing 0.01% cysteamine (which is sufficient to maintain active chronic duodenal ulcers) (9). Also from the second day (i.e., when the acute ulcers were formed), certain groups of rats began to receive daily doses of lergotrile or bromocriptine. It can be seen that after 7 days of treatment only the intensity of duodenal ulcers was significantly decreased while after 21 days both the incidence and intensity of lesions were markedly reduced (Table III). These data thus suggest an accelerated healing of chronic duodenal ulcers in rats treated with these ergots.

More recently, we initiated a series of experiments to test the activity of new derivatives of apomorphine, some of which had been classified as dopamine agonists or antagonists (40-45). Pretreatment of rats with 50 or 100 μg/100 g body weight of (-)N-n-propylnorapomorphine (NPA) or the prodrug (-) 10,11-methylenedioxy-N-n-propyl noraporphine (MDO-NPA) (43) once daily for 7 days significantly, by 50% or more, (except NPA at 50 μg dose) diminished the intensity of cysteamine-induced duodenal ulcers (Table IV). Acute administration of these two aporphine derivatives (i.e., on the day of cysteamine treatment) did not modify the duodenal ulceration but it - like chronic doses - reduced gastric acid output (46). On the other hand, even the acute doses (50 or 100 μg/100 g of (-)N-(chloroethyl) norapomorphine (NCA) or (+)butaclamol at the same dose range aggravated the intensity of duodenal ulcers and mortality in cysteamine-treated rats (Table IV). To our surprise, in very recent experiments, daily pretreatment for 1 week with NCA resulted in slight but statistically significant protection against acute duodenal ulcers induced by cysteamine. These biphasic results, nevertheless, might be similar to the two-site actions of NCA recently described by Lehmann and Langer (47), i.e., NCA appeared to be a reversible dopamine agonist at the dopamine autoreceptors, while acting as an irreversible antagonist of postsynaptic dopamine receptors. Thus, further studies are needed to characterize the site and extent of involvement of dopamine receptors in the pathogenesis of duodenal ulceration and to identify potent dopamine-related drugs which may modify this ulcer disease only in a beneficial way (e.g., necessary for prevention or treatment).

TABLE III. Effect of Post-treatment with Lergotrile or
Bromocriptine on the Cysteamine-induced Chronic Duodenal Ulcer
In the Rat

Group	Pretreatment[#]	Duration of Experiment	Duodenal ulcer Incidence (Pos./Tot.)	Intensity (Scale:0-3)	Mortality (%)
1.	Water	7-21 days	38/45	2.0 ± 0.4	47
2.	Lergotrile	7 days	19/25	1.0 ± 0.1*	20
3.	Bromocriptine	7 days	19/25	1.1 ± 0.2*	20
4.	Lergotrile	21 days	9/20*	0.6 ± 0.2*	45
5.	Bromocriptine	21 days	8/21*	1.1 ± 0.2*	33

[#] In addition, rats of all groups received cysteamine.

* = $p < 0.005$

(Reprinted from Ref. No. 59)

TABLE IV. Effect of MDO-NPA, NPA, NCA or (+)butaclamol
on Cysteamine-induced Duodenal Ulcer in the Rat

Group	Pretreatment	Dose (µg/ 100g)	Duodenal ulcer	
			Incidence (Pos./Tot.) (%)	Intensity (Scale:0-3)
1.	Control	—	10/12 83	1.8+0.2
2.	MDO-NPA	50	4/6 67	0.8+0.2 *
3.	MDO-NPA	100	3/6 50	0.5+0.1 +
4.	NPA	50	8/9 89	1.1+0.3
5.	NPA	100	6/9 67	0.9+0.1 *
6.	NCA	50	5/6 83	2.0+0.3
7.	NCA	100	6/6 100	2.4+0.1 *
8.	(+)Butaclamol	50	6/6 100	2.6+0.2 *
9.	(+)Butaclamol	100	6/6 100	2.4+0.1 *

The groups consisted of 3-4 Sprague-Dawley female rats (160-180g).
Each experiment was repeated at least twice and the results of
those groups were pooled. The dopamine agonists MDO-NPA and NPA
were injected sc once daily for seven days prior to the
administration of cysteamine HCl (Aldrich) 28 mg/100 g po three
times with 3 h intervals. The dopamine antagonists NCA and
(+)butaclamol were injected sc three times, 30 min before each
dose of cysteamine. The animals were killed 48 h after the
duodenal ulcerogen. The intensity of duodenal ulcer was evaluated
on a scale of 0-3, where 0 = no ulcer, 1 = superficial mucosal
erosion, 2 = transmural necrosis, deep ulcer, 3 = perforated or
penetrated duodenal ulcer.

* = p<0.05; + = p<0.005

Biochemical Studies with Catecholamines

In this section we shall review our results only with
catecholamines although extensive biochemical studies were
performed with serotonin, GABA, somatostatin and sulfhydryl-
containing agents as well (34,35,48,49). At this point it must
be emphasized that we do not claim dopamine derangements to be
the only major problem in duodenal ulceration. Rather, dopamine
may play a hitherto unrecognized role in the chain-of-events
leading to the multifactoral formation of duodenal ulcer. What
may, nevertheless, distinguish dopamine from other elements in
this pathogenetic sequence is that like histamine, dopamine also
appears amenable to pharmacologic modulation, but unlike
histamine, this catecholamine may have both peripheral and
central roles in causation of duodenal ulceration (4,34,37,49).
Concentrations of dopamine and norepinephrine were measured
in the entire brain and seven brain regions (50), adrenals,
forestomach, glandular stomach, proximal duodenum, mucosa of
glandular stomach and upper duodenum, as well as pancreas of rats
in dose- and time-response experiments after the administration
of cysteamine or propionitrile (4,34). Norepinephrine levels
were decreased to various extents in all of the tissues studied.
Dopamine concentrations, on the other hand, either did not change
or increased, e.g., prominent and transient elevations in the
cortex and midbrain and adrenals (30 min - 1 h after the duodenal
ulcerogens), or prolonged increase in the hypothalamus, medulla
oblongata and glandular stomach (30 min - 8 h). These changes
might be correlated with decreased tissue somatostatin, elevated
serum gastrin levels and enhanced gastric acid output detectable
in these time intervals (4,29,31,34,35,38). Dopamine metabolite
(DOPAC and HVA) levels were also increased and these changes
could not be explained by the inhibition of
dopamine-β-hydroxylase (51).
More recently, these studies were expanded by the
measurements of dopamine and norepinephrine turnover after the
inhibition of tyrosine hydroxylase by α-methyl-p-tyrosine (52)
which indeed caused a time-dependent depletion of brain dopamine
and nonadrenaline levels leading to marked aggravation of
cysteamine-induced duodenal ulcer and mortality (37). When
cysteamine or propionitrile were given to rats after pretreatment
with α-methyl-p-tyrosine, the turnover of nonepinephrine was
either unaffected or accelerated in certain brain regions, while
that of dopamine was retarded at virtually all of the time
intervals studied. This effect was especially prominent in the
corpus striatum, midbrain and medulla oblongata (cf., origin of
the vagus nerve). Thus, these results also support the notion
that dopamine synthesis, turnover and/or binding sites may be
selectively affected during experimental duodenal ulceration.
Changes seen with norepinephrine (e.g., increased turnover and
decreased levels) are identical to those seen in stress reaction

or after the administration of large amounts of virtually any drug which, on the other hand, does not affect dopamine levels and turnover (4,53).

Evidence for the involvement of dopamine in duodenal ulceration is also accumulating from ongoing radioligand binding studies. For these experiments [³H]-dopamine and [³H]-haloperidol are being used with slight modification of binding assays (54,55). With these techniques we were able to demonstrate saturable binding sites for [³H]-dopamine in gastric and duodenal mucosal and muscularis propria membranes (56,57). The order of specificity of displacing ligands was dopamine> norepinephrine>isoproterenol. Scatchard analysis showed a uniform population of binding sites with a maximum binding (B_{max}) twice higher (about 28 pmol/mg proteins) in the muscular layers than in the mucosa (about 14 pmol/mg proteins) of either stomach or duodenum (Figures 5,6). Other experiments revealed the presence of a dopamine-sensitive adenylate cyclase in duodenal and gastric mucosa (58). The activity of the enzyme was increased three-fold by dopamine and slightly enhanced by norepinephrine or histamine in both in vivo and in vitro experiments (Table V).

The duodenal ulcerogens also influence these newly demonstrated dopamine binding sites in the stomach and duodenum. In vitro only cysteamine, and not the nitrile-[³H]-containing duodenal ulcerogens (e.g., propionitrile, n-butyronitrile) displaced [³H]-dopamine from duodenal mucosal membranes. In vivo, however, either cysteamine or propionitrile administration to rats had a time-dependent action on [³H]-dopamine binding sites (Table VI). After an initial (1 h) 50% (approximately) decrease in binding to muscularis propria, an increase of the same magnitude to mucosal membranes was seen 4 h after the administration of duodenal ulcerogens (Table VI). These alterations are in correlation with the early motility and late secretory changes observed in the course of duodenal ulceration (4,34,57,59)

Correlations and Implications

The results reviewed here strongly suggest a dopamine disorder in the pathogenesis of experimental, chemically induced duodenal ulcer. Since the cysteamine- or propionitrile-induced ulcer in the rat strongly resembles the human duodenal ulcer disease in functional and morphologic criteria (7-10), the extrapolations concerning dopamine probably can be expanded to etiologic considerations in man (more direct evidence through the possible connection of peptic ulcer disease and dopamine-dependent or dopamine-sensitive disorders will be discussed later).

The first suggestion of the duodenal ulcer-dopamine connection surfaced with the recognition that the duodenal

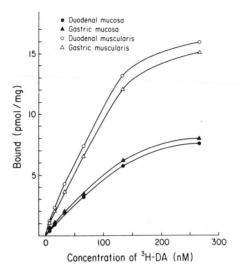

Figure 5. Saturability of specific ³H-dopamine binding to gastric and duodenal membranes. (Reproduced with permission from Ref. 57. Copyright 1982, Pergamon Press Ltd.)

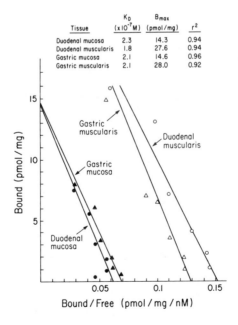

Figure 6. Scatchard analysis of specific ³H-dopamine binding to gastric and duodenal membranes. (Reproduced with permission from Ref. 57. Copyright 1982, Pergamon Press Ltd.)

TABLE V. Effect of Dopamine, Noradrenaline or Histamine on the
Activity of Adenylate Cyclase of Rat Duodenal Mucosa
in vitro and in vivo

Treatment	In vitro (A) Adenylate cyclase Specific activity(C)(%)		In vivo (B) Adenylate cyclase Specific activity(%)	
None	141.9 + 35.0	100	97.3 + 11.3	100
Dopamine	425.4 + 70.5	300	307.1 + 151.2	316
Noradrenaline	247.9 + 45.0	175		
Histamine	200.1 + 77.2	141	234.8 + 96.4	241

Adenylate cyclase activity was measured in mucosal homogenates by
radioimmunoassay of the liberated cyclic AMP.

(A) In vitro treatment: 50µl of 10^{-3}M solultion of monoamines was
added to the incubation medium.
(B) In vivo treatment: rats were injected sc with dopamine,
5mg/100g, or histamine, 5mg/100g, 20 min before sacrafice.
(C) Specific activity: cyclic AMP pmol/mg proteins/min.

(Reprinted from Ref. No. 57)

TABLE VI. Effect of Cysteamine or Propionitrile on Specific
[3]H-Dopamine Binding to Duodenal and Gastric Membranes in the Rat

Tissues	^{3}H-dopamine (pmol/mg protein)					
	Control		1 hr		4 hr	8 hr
	Cyst.	Prop.	Cyst.	Prop.	Cyst.	Prop.
Du.mu	2.9+0.2	2.5+0.9	2.7+1.4	4.9+0.7*	5.9+1.0*	2.8+0.3 4.5+1.1
Du.ms	7.7+0.8	2.7+0.5#	2.1+0.6#	7.2+0.7	6.1+0.2	12.4+2.5 20.1+6.7
Gs.mu	2.4+0.2	2.0+0.5	2.6+0.9	4.4+0.5#	5.3+0.8*	2.6+0.8 1.9+0.3
Gs.ms	7.7+0.7	3.3+0.8#	8.7+0.3	4.6+0.4	5.9+0.1	6.7+1.5 9.4+2.5

* = $p < 0.01$; # = $p < 0.001$
Cyst. = Cysteamine; Prop. = Propionitrile; Du.mu = Duodenal
mucosa; Du.ms = Duodenal muscle (muscularis propria); Gs.mu =
Gastric mucosa; Gs.ms = Gastric muscle

(Reprinted from Ref. No. 57)

ulcerogenic action - although surprising because of the previous
lack of agents to produce experimental duodenal ulcer - is not
limited to propionitrile (5), cysteamine (6), and the few aryl
chemicals (4,18-20) initially discovered to cause duodenal ulcer
in the rat. More precisely, a structure-activity correlation and
extrapolation was recognized indicating that the duodenal
ulcerogenic action of chemicals seems to be associated with
"two-carbon" structures bearing reactive nucleophilic radicals
(e.g., -SH, -CN, -NH$_2$) on one or both ends of the 2-C moiety
(4,20-22). Quantitative structure-activity relations (QSAR),
however, also indicated that two or more groups of chemicals
capable of causing duodenal ulcer might exist (22). It also
became evident that other factors or properties of the radicals
(e.g., molar refractivity, electronic parameters) are also
important correlates or prerequisites for the duodenal
ulcerogenic and/or adrenocorticolytic action of alkyl or aryl
chemicals. These correlations have been extensively discussed in
other recent publications (4,22).

The most important implications of structure-activity
studies were probably not only the ability to predict the
duodenal ulcerogenic action of exogenous chemicals, but also (or
rather) the possibility to consider endogenous putative mediators
or modulators of duodenal ulcerations. Among endogenous
ethylamines, histamine has long been known to be the most potent
gastric secretogogue, while dopamine and serotonin had not been
considered until recently in the etiologic role of duodenal
ulceration. We chose to investigate in detail the possible role
of dopamine in this very frequent gut disorder. The first,
pharmacologic approach proved to be rewarding because we selected
a few dopamine agonists (e.g., bromocriptine, lergotrile,
apomorphine, L-DOPA) and antagonists (e.g., haloperidol,
pimozide) capable of preventing and aggravating, respectively,
the experimental duodenal ulcers (36). The second task, the
pharmacologic and biochemical designation of this action of
dopamine in the gut is difficult because of the new and evolving
criteria for designation of dopamine actions and receptors
(60-63). However, because of the similarity of action elicited
by duodenal ulcerogens in the gut (e.g., vasodilation and
increased blood flow), the recently demonstrated dopamine binding
sites in the stomach and duodenum seem to be similar to dopamine
vascular receptors described for the renal artery (60). Because
of the association with adenylate cyclase, they may also be
designated D-1 receptors (61).

Possible functional correlates of the apparent biphasic or
multiple-site action of ergot alkaloids are the contradictory
reports on the effect of bromocriptine on gastric acid secretion
(64-66). Other clues are the possible α-adrenergic and
anticholinergic actions of ergot alkaloids, and those effects may
at least explain the anti-secretory influence of a single dose
but not the complete duodenal anti-ulcerogenic action most
prominent after prolonged treatment with these drugs.

The availability of a new series of apomorphine derivatives
allowed more extensive manipulation of drugs to prevent and treat
experimental duodenal ulcer disease. Indeed, the recently tested
NPA and MDO-NPA are about 10 and 200 more potent than ergots and
H_2 receptor antagonists, respectively, but they are comparatively
less effective because even at the most optimal dose these
apomorphines produced at the most slightly more than 50%
reduction in the intensity of cysteamine-induced duodenal ulcers
(46). In these comparisons one also has to stress the
availability of oral route of administration for MDO-NPA which is
the only apomorphine derivative not requiring parenteral
injection (43). Furthermore, the biphasic or antagonistic action
of NCA in experimental duodenal ulceration (e.g., aggravating the
ulcers in acute experiments but exerting mild protective action
after chronic treatment) and the differential effect of this
compound on putative dopamine autoreceptors and postsynaptic
receptors (47) also call for a new quest for novel drug
derivatives, especially dopaminergic agonists to be evaluated for
effectiveness in duodenal ulceration. Although the best
dopamine-related pharmacologic approach to duodenal ulcer
prevention and treatment remains to be investigated, it is
reassuring to know that endogenous precursors of dopamine (e.g.,
L-tyrosine or L-DOPA) are also effective in the prevention of
cysteamine-induced duodenal ulcers (67).
     The recently demonstrated binding sites for dopamine in
gastric and duodenal mucosa and muscularis propria add new
emphasis to the pathophysiologic and pharmacologic role of
dopamine in duodenal ulceration (56,57). It has long been known
that gastric motility disorders are amenable to modulation by
certain dopamine antagonists and it has recently been reported
that infusion of dopamine reduces the pentagastrin-stimulated
gastric secretion in man (68). These pharmacologic manipulations
in the gut have offered new biochemical explanations; it remains
to be elucidated what are the implications of having twice as
many dopamine binding sites in the muscles of the stomach and
duodenum as in the mucosa of these organs. The duodenum has long
been known to contain relatively high concentrations of dopamine
(69,70); one of the physiologic and pathologic roles for this
catecholamine could well be connected to ulcer disease. Dopamine
may have multiple sites of action relevant to duodenal ulceration
even in peripheral organs, e.g., stimulation of secretion of
pancreatic bicarbonate (68,71) which contributes to
neutralization of acid, inhibition of gastric acid output (36,59)
and regulation of gastric and duodenal motility which, according
to our preliminary results, may have a major role in the
development, and prevention of duodenal ulcer (cf., lack of
correlation between duodenal anti-ulcerogenic and gastric
anti-secretory actions of ergot alkaloids). The central general
inhibitory role of dopamine (injected icv to rats) and its

specific duodenal anti-ulcerogenic action (37) are also well documented.
Circumstantial evidence directly implicating dopamine in the pathogenesis of duodenal ulcer in man is the unusual incidence of peptic ulcer disease in dopamine-deficient disorders. From purely descriptive clinical and epidemiologic studies we know that patients with Parkinson's disease, before the introduction of dopamine therapy, had an excess of ulcer disease (72). One report even comments on the curiosity that after initiation of L-DOPA administration the ulcer symptoms have virtually disappeared (72). On the other hand, less clearly, schizophrenia which is associated with dopamine excess and/or receptor hyperactivity is accompanied by virtual lack, or decreased prevalence, of peptic ulcer (73-76). Schizophrenia associated with ulcer disease has been viewed as a reportable curiosity in medical literature (75). At present, possibly because of the widespread therapeutic application of neuroleptics, the lack of peptic ulcer disease in schizophrenics is less striking than in the past. On the other hand, we recently observed in our autopsy series perforated duodenal ulcers in two schizophrenic patients who had been on large doses of haloperidol therapy (Szabo, unpublished observation). Thus, even in man, dopamine may indeed be implicated in the pathogenesis of duodenal ulcer disease.

Literature Cited

1. Sleisenger, M.H., Fordtran, J.S. "Gastrointestinal Disease"; W.B. Saunders Company: Philadelphia, 1978.
2. Wormsley, K.G. "Duodenal Ulcer"; A. Wheaton & Co., Ltd.: Exeter, 1979.
3. Robert, A., Szabo, S. "Selye's Guide to Stress Research"; Selye, H., Ed.; Van Nostrand: New York, in press.
4. Szabo, S., Horner, H.C., Gallagher, G.T. "Drugs and Peptic Ulcer"; Pfeiffer, C.J., Ed.; CRC Press, Inc.: Boca Raton, 1981; p. 55-74.
5. Szabo, S.; Selye, H. Arch. Pathol., 1972, 93, 389.
6. Selye, H.; Szabo, S. Nature, 1973, 244, 458.
7. Robert, A.; Nezamis, J.E.; Lancaster, C.; Badalamenti, J.N. Digestion, 1974, 11, 199-214.
8. Ishii, Y.; Fujii, Y.; Homma, M. Eur. J. Pharmacol., 1976, 36, 331-336.
9. Szabo, S. Amer. J. Pathol, 1978, 93, 273-276.
10. Szabo, S.; Haith, L.R., Jr.; Reynolds, E.S. Amer. J. Dig. Dis., 1979, 24, 471-477.
11. Bristol, J.A.; Kaminski, J.J. Ann. Rep. Med. Chem., 1982, 17, 89-98.
12. Robert, A. Biol. Gastroenterol. (Paris), 1974, 7, 145-161.
13. Seronde, J., Jr. "Peptic Ulcer"; Pfeiffer, C.J., Ed.; Lippincott; Philadelphia, 1971; p. 3-12.
14. Robert, A.; Stout, D.J.; Dale, J.E. Gastroenterology, 1970, 59, 95-102.

15. Okabe, S., Pfeiffer, C.J. "Peptic Ulcer"; Pfeiffer, C.J.,
     Ed.; Lippincott; Philadelphia, Toronto, 1971; p. 13-20.
16. Szabo, S. Amer. J. Pathol., 1978, 93, 273-276.
17. McComb, D.J.; Kovacs, K.; Horner, H.C.; Gallagher, G.T.;
     Schwedes, U.; Usadel, K.H.; Szabo, S. Exp. Mol. Pathol.,
     1981, 35, 422-434.
18. Szabo, S. "Hormones and Resistance"; Selye, H., Ed.;
     Springer; New York, 1971; p. 176.
19. Selye, H. Proc. Soc. Exp. Biol. Med., 1973, 142, 1192-1194.
20. Szabo, S. Proc. 5th World Congr. Gastroenterol. (Mexico
     City), 1974, 169.
21. Szabo, S.; Reynolds, E.S. Environ. Health Perspect., 1975,
     11, 135-140.
22. Szabo, S.; Reynolds, E.S.; Unger, S.H. J. Pharmacol. Exp.
     Ther., 1982, 223, 68-76.
23. Black, J.; Duncan, W.; Emmett, J.; Ganellin, C.R.; Hesselbo,
     T.; Parsons, M.; Wyllie, J. Agents Actions, 1973, 3,
     133-137.
24. Brimblecombe, R.W.; Duncan, W.A.M.; Durant, C.J.; Emmett,
     J.C.; Ganellin, C.R.; Parsons, M.E. J. Int. Med. Res., 1975,
     3, 86-92.
25. Robert, A.; Nezamis, J.E.; Lancaster, E. Toxicol. Appl.
     Pharmacol., 1975, 31, 201-207.
26. Groves, W.G.; Schlosser, J.H.; Mead, F.D. Res. Commun. Chem.
     Pathol. Pharmacol., 1974, 9, 523-534.
27. Ravokatra, A.; Loiseau, A.; Ratsimamanga-Urverg, S.; Nigeon-
     Dureuil, M.; Ratsimamanga, A.R. C.R. Acad. Sci., 1974, Ser.
     D, 278, 2317.
28. Borella, L.E., Suthaler, K., Lippmann, W. "Progress in
     Peptic Ulcer"; Mozsik, Gy, Javor, T., Eds.; Akademiai Kiado;
     Budapest, 1976; p. 585-596.
29. Kirkegaard, P.; Poulsen, S.S.; Loud, F.B.; Halse, C.;
     Christiansen, J. Scand. J. Gaastroent., 1980, 15, 621-624.
30. Poulsen, S.S.; Kirkegaard, P.; Skov Olsen, P.; Christiansen,
     J. Scand. J. Gastroent., 1981, 16, 459-464.
31. Kirkegaard, P.; Petersen, B.; Skov Olsen, P.; Poulsen, S.S.;
     Christiansen, J. Scand. J. Gastroent., 1982, 17, 609-612.
32. Tsunoda, S.; Yabana, T. Sapporo Med. J., 1980, 49, 281-302.
33. Leithold, M.; Englehorn, R.; Schierok, H.J.; Seidel, H.;
     Eliasson, K. Therapiewoche, 1977, 27, 1532.
34. Szabo, S. Stress, 1980, 1(2), 25-36.
35. Szabo, S.; Reichlin, S. Endocrinology, 1981, 109, 2255-2257
36. Szabo, S. Lancet, 1979, 2, 880-882.
37. Horner, H.C.; Szabo, S. Life Sci., 1981, 29, 2437-2443.
38. Gallagher, G.T.; Szabo, S. Fed. Proc., 1980, 39, 326.
39. Adler, R.S.; Gallagher, G.T.; Szabo, S. Fed. Proc., 1981,
     40, 512.
40. Costall, B.; Fortune, D.H.; Law, S.J.; Naylor, R.J.;
     Neumeyer, J.L.; Nohria, V. Nature, 1980, 285, 571-573.

41. Neumeyer, J.L.; Arana, G.W.; Law, S.J.; Lamont, J.S.; Kula, N.S.; Baldessarini, R.J. J. Med. Chem., 1981, 24, 1440-1445
42. Neumeyer, J.L.; Arana, G.W.; Ram, V.J.; Kula, N.S.; Baldessarini, R.J. J. Med. Chem., 1982, 25, 990-992.
43. Baldessarini, R.J.; Neumeyer, J.L.; Campbell, A.; Sperk, G.; Ram, V.J.; Arana, G.W.; Kula, N.S. Eur. J. Pharm., 1982, 77 87-88.
44. Arana, G.W.; Baldessarini, R.J.; Neumeyer, J.L. Acta. Pharm Suec., 1982 (in press).
45. Neumeyer, J.L.; Arana, G.W.; Ram, V.J.; Baldessarini, R.J. Acta. Pharm. Suec., 1982 (in press).
46. Neumeyer, J.L.; Szabo, S. Nature, 1982 (submitted).
47. Lehmann, J.; Langer, S.Z. Eur. J. Pharm., 1982, 77, 85-86.
48. Szabo, S., Trier, J.S., Gallagher, G.T., Frankel, P.W. "Basic Mechanisms of Gastrointestinal Mucosal Cell Injury and Protection"; Harmon, J.W., Ed.; Williams & Wilkins; Baltimore, 1981; p. 249-261.
49. Szabo, S., Horner, H.C., Maull, E.A. "Problems in GABA Research From Brain to Bacteria"; Okada, Y., Roberts, E., Eds.; Excerpta Medica; Amsterdam, 1982; p. 147-155.
50. Glowinski, J.; Iversen, L.L. J. Neurochem., 1966, 13, 655-669.
51. Horner, H.C.; Szabo, S. Gastroenterology, 1979, 76, 1305.
52. Brodie, B.B.; Costa, E.; Dlabac, A.; Neff, N.H.; Smookler, H.H. J. Pharmacol. Exp. Ther., 1966, 154, 493-498.
53. Selye, H. "Hormones and Resistance"; Springer; New York, 1971.
54. Burt, D.R.; Creese, I.; Snyder, S.H. Mol. Pharmacol., 1976, 12, 800-812.
55. List, L; Titeler, M.; Seeman, P. Biochem. Pharmacol., 1980, 29, 1621-1622.
56. Sandrock, A. Gastroenterology, 1981, 80, 1362.
57. Szabo, S., Sandrock, A.W., Nafradi, J., Maull, E.A., Gallagher, G.T., Blyzniuk, A. "Advances in Dopamine Research"; Kohsaka, M. et al., Eds.; Pergamon Press; Oxford, 1982; p. 165-170.
58. Nafradi, J.; Szabo, S. Fed. Proc., 1981, 40, 709.
59. Szabo, S., Gallagher, G.T., Blyzniuk, A., Maull, E.A., Sandrock, A.W. "Advances in Pharmacology and Therapeutics II"; Vol. 5; Yoshida, H., Hagihara, Y., Ebashi, S., Eds.; Pergamon Press; Oxford, 1982; p. 263-268.
60. Goldberg, L.I.; Volkman, P.H.; Kohli, J.D. Annu. Rev. Pharmacol. Toxicol., 1978, 18, 57-79.
61. Kebabian, J.W.; Calne, D.B. Nature, 1979, 277, 93-96.
62. Seeman, P. Pharmacol. Rev., 1981, 32, 229-313.
63. Creese, I.; Sibley, D.R. Biochem. Pharmacol., 1982, 31, 2568-2569.
64. Hirst, B.H.; Reed, J.D.; Gomez-Pan, A.; Albert, L. Clin. Endocrinol., 1976, 5, 723-729.
65. Caldara, R.; Grimaldi, D.; Ferrari, C. Lancet, 1977, 1, 902-903.

66. Reding, P.; De Graef, J.; Barbier, P. Lancet, 1978, 1, 1202-1203.
67. Oishi, T.; Szabo, S. Fed. Proc., 1982, 41, 1719.
68. Valenzuela, J.E.; Defilippi, C.; Diaz, H.; Navia, E.; Mueno, Y. Gastroenterology, 1979, 76, 323-326.
69. Landsberg, L.; Berardino, M.B.; Silva, P. Biochem. Pharmacol., 1975, 24, 1167-1174.
70. Christensen, N.J.; Brandsborg, O. J. Clin. Lab. Invest., 1974, 34, 315-320.
71. Hashimoto, K., Furuta, Y., Iwatsuki, K. "Frontiers in Catecholamine Research"' Usdin, E., Snyder, S.H., Eds.; Pergamon Press; New York, 1973; p. 825-829.
72. Strang, R.R. Med. J. Australia, 1965, 1, 842-843.
73. Pollak, O.J.; Kreplick, F. J. Nerv. Ment. Dis., 1945, 101, 1-8.
74. Gosling, R.H. J. Psychosom. Res., 1958, 2, 285-301.
75. Samet, E.T.; White, M.S.; Vaughn, A.M. Amer. J. Dig. Dis., 1957, 2, 437-441.
76. Hinterhuber, H.; Hochenegg, L. Arch. Psychiat. Nervenkr., 1975, 220, 335-345.

RECEIVED March 17, 1983

Commentary: Dopamine Agonists and Antagonists in Duodenal Ulcer Disease

GEORGE W. ARANA

Harvard Medical School and the Mailman Research Center, Department of
Psychiatry, McLean Affiliate of the Massachusetts General Hospital,
Belmont, MA 02178

The presentation by Drs. Szabo and Neumeyer summarizes recent
investigations which have explored the possible role of dopamine
(DA) in peptic ulcer disease. As is true for other medical conditions, including such common illnesses as hypertension, arthri-
tis, and dementia, peptic ulcer probably represents a dysfunction
of several physiological systems having a final common pathway of
expression. In ulcer disease, this expression is erosion and in
severe cases, perforation of the gastric or duodenal wall. Szabo
has been instrumental in the development of an animal model for
duodenal ulcer disease which involves injection of either cysteamine or propionitrile. These are small, dicarbon molecules with
an amino group and sulfhydryl, or cyano substituents, respectively. Using this toxin-induced experimental model of ulcer in rats,
Szabo has reported that solitary, sometimes perforating, ulcers
are reliably and consistently produced on the anterior duodenum.
More recently, he found that bromocriptine and lergotrile, ergoline derivatives with DA agonist properties, reduce the incidence
of ulcerogenesis in this animal model. Subsequently, the authors
undertook studies to investigate further possible links between
ulcerogenesis and DA receptors. Thus, they evaluated the effects
of aporphine analogs with potent DA-agonist actions upon ulcer
formation in rats. Following seven days of treatment (1 mg/kg/
day) with either N(n-propyl)norapomorphine (NPA) or methylene-
dioxy-N(n-propyl)norapomorphine (MDO-NPA), there was a 3-fold and
2-fold reduction, respectively, in ulcer production. Similar
findings obtained when either bromocriptine or lergotrile were
administered at double the dose (2 mg/kg/day) suggest high
potency for the aporphine compounds. These findings are exciting
and deserve further exploration, particularly since they suggest
a role for DA in the pathophysiology of ulcer disease.

When animals were pretreated with a potent DA blocking agent
(haloperidol, pimozide, or (+)-butaclamol), there was a significant increase of ulcerogenesis and associated mortality in experimental animals. Although this effect was not as large as the

0097–6156/83/0224–0197$06.00/0

reduction of ulcer production seen with DA agonists (2-3-fold effect), this result may reflect the ceiling imposed by the control condition (at 60% of the maximal effect). The increase in mortality, however, was 3-fold; here, the baseline for mortality was at 20% of maximal effect and hence a 500% increase was possible. If the baseline for ulcerogenesis could be lowered by manipulation of the experimental design, the enhancement of ulcerogenesis with DA blocking agents might be even more clearly demonstrated. Nevertheless, there is a strong suggestion that DA agonists and antagonists affect cysteamine and propionitrile-induced ulcer production.

A useful area for investigation may be more complete study of the effects of other DA agonists such as DA, epinine, ADTN, APO, NPA and other aporphine analogs as well as DA antagonists such as phenothiazines, butyrophenones, and benzamides. It also may be useful to study a series of ergoline compounds. It may become clearer after a larger series of compounds are screened, what neuroregulatory system is most relevant to pursue, and similarly, which ^{3}H-ligand would be the most useful for receptor characterizations. For instance, the importance of a two-carbon moiety could be tested by administering tyramine or tryptamine to experimental animals; these compounds have ring structures and ethylamine moieties similar to DA, but are sufficiently dissimilar that critical differences may be demonstrated by their effects or lack of effects. Similarly, the rank-potencies of various DA agonists or antagonists may indicate clear structure-activity relationships of the cysteamine and propionitrile-induced ulcerogenesis. For example, a study of ergolines may help to test the importance of the tetracyclic ring of ergot alkaloids rather than the ethylamine moiety which is buried in the ring structure. Thus, pharmacologic characterization may elucidate the role of DA in the pathophysiology of toxin-induced ulcers in the rat. Binding studies using radioactively labeled compounds can be misleading unless pharmacologic characterization is extensively pursued. Thus, although the binding studies reported by Szabo and Neumeyer are intriguing, the kinetics of ^{3}H-DA binding involved Scatchard analyses that revealed appreciable curvature, suggesting multiple binding sites. The inflection point of the saturation curves shown (Figure 5) are well below the cited B_{max}s (for mucosa, saturation curve suggests B_{max} = 7-9 pmol/mg cited as 14.3 and 14.6 pmol/mg in Figure 6; for muscularis saturation curve suggests B_{max} = 16-18 pmol/mg [Figure 6 cites 27.6 and 28.0 pmol/mg]). Competition for the binding sites by cysteamine or propionitrile was evidently not evaluated. Thus it is not clear whether residual amounts of these compounds in tissue preparations may have interfered with the binding (Table 6) especially at early times post-injection. In fact, the marked decrease in ^{3}H-DA binding to muscularis at one hour after injection of cysteamine or propionitrile may reflect such interference, and may suggest that these dicarbon molecules have some affinity for a dopamine receptor labeled by ^{3}H-DA in both duodenal and gastric tissues.

In conclusion, these are interesting and exciting findings which may point to new and less toxic treatments of ulcer disease.

RECEIVED February 3, 1983

The Development of Novel Dopamine Agonists

DAVID E. NICHOLS

Purdue University, Department of Medicinal Chemistry and Pharmacognosy,
School of Pharmacy and Pharmacal Sciences, West Lafayette, IN 47907

Some of the significant refinements in the
knowledge of structure—activity relationships for
dopamine agonists which have been made over the
past several years are presented. Selected
examples of structural dissection and rigid
analogue design are discussed. Possible binding
orientations for several types of dopamine
agonists, including the ergolines, are examined
using stereopair superpositions of framework
molecular models. The postulated binding
orientation for ergolines allows an explanation to
be made for the increased dopaminergic activity of
13-hydroxylated ergoline metabolites. Emphasis is
added to a discussion of the need for studies to
evaluate the possible importance of the nitrogen
electron pair orientation in determining dopamine
receptor selectivity.

One can view the "rational" development of drug leads as
proceeding along several possible directions. In particular,
approaches such as structural dissection of complex natural
products, or rigidification of a natural substrate, hormone, or
pharmacophore by incorporation into a more complex structure
have been useful. Quantitative methods are useful for
optimizing activity but seldom give rise to novel leads. The
development of peripheral dopamine agonists exemplifies the
processes of structural dissection and rigidification and serves
as a good example of drug design and lead development.

To carry out design it is required, by definition, that an
end point be identified. That is, what is the therapeutic goal?
Or, on the molecular level, what is the target for the drug?
The confusion over dopamine receptor subtypes will be commented
on by others at this symposium but represents a serious obstacle
to design of selective dopamine agonists. A focus on
peripheral agonists simplifies matters, for at least in this

0097–6156/83/0224–0201$06.00/0

case one has physiological endpoints which are relatively easy to quantitate. The situation in the central nervous system, at best, is needlessly complex at the present time, with four and perhaps even five different dopamine binding <u>sites</u> having been postulated. This discussion will deal mainly with compounds which interact with peripheral D-1, DA_1, D-2 or DA_2 receptors. Nevertheless, one should bear in mind that there ought to be some functional and topographical similarities between central and peripheral receptors. This is simply based on the idea that different subtypes must have originated through a common receptor "progenitor" with evolution leading mainly to changes in auxilliary binding sites and the immediate environment around the active site.

One additional problem in agonist development to be aware of is the difficulty of designing agonists, as compared with antagonists. Gund (<u>1</u>) has noted that agonists, which must be accommodated to the delicate energetic balance of a membrane recycling process, are more difficult to design than antagonists, which may bind to any accessible conformation of the receptor.

Structural Dissection

Some of the earliest and most complete efforts at structural dissection have been carried through by Cannon and his co-workers at the University of Iowa. These studies were initially directed toward elucidation of the pharmacophoric element within the structure of the emetic agent apomorphine, I. The suggestion by Pinder et al. (<u>2</u>) that the 5,6-dihydroxy-2-aminotetralin fragment was the active moiety was followed in short order by the report of Cannon et al. (<u>3</u>) that the N,N-dimethyl derivative ("M-7") II was a potent emetic in the dog. Additional pharmacology on M-7 provided by Long et al. (<u>4</u>) further illustrated the similarity between I and II. Both

I   II

compounds block the increase in heart rate produced by
postganglionic stimulation of the cardioaccelerator nerve.
Their action is blocked by dopamine receptor antagonists such as
chlorpromazine and haloperidol. This action was attributed to
an interaction with inhibitory dopamine receptors located on
adrenergic nerve terminals, which would now be classified as DA_2
receptors.

Further structural dissection has shown that the essential
requirement for an apomorphine-like action in the periphery is a
relatively rigid fragment where the amino group is tertiary and
is held approximately in the plane of the aromatic ring. Within
the 2-amino-1,2,3,4-tetrahydronaphthalenes, it is the 5-hydroxy
which is essential for emetic and DA_2 activity. When the
reduced ring is contracted to yield a 2-aminoindan such as III,
substantial activity is still observed (5). Thus, the view of

III

the receptor with a binding site for a hydroxy oriented "meta"
with respect to the side chain, should be taken to mean that the
amino group is separated in space from the hydroxy by 6.5-7.5 Å.
The receptor undoubtedly possesses the flexibility to allow
accommodation to a variety of agonist conformations. This
latter is a point which is frequently overlooked and cannot be
overemphasized. Often it seems that drug-receptor interactions
are viewed from a narrow "lock-and-key" approach.

A second important area where structure dissection has been
applied is in the study of dopaminergic ergolines. The potent
dopamine-like action of the ergolines was first noted in the
early 1970's (6,7). However, ergolines deviate so obviously in
structure from other classes of dopaminergic agents that it is
difficult to visualize which portions of the molecule represent
the pharmacophore. In 1976 this author proposed a structural
similarity between apomorphine and the ergolines which
explicitly noted the importance of the pyrrole ring in the
ergolines as corresponding to the catechol ring of more
conventional agonists (8). Superposition of the pyrrole and
catechol rings, the basic nitrogen atoms, and chiral centers C5
and C6a in the ergolines and apomorphine, respectively, is

illustrated in Figure 1a where the ergoline, shown as the dashed structure, is pergolide and the solid structure is apomorphine. Each is shown as the more active R enantiomer. Figure 1b is an edge-on view, where it is seen that, with the exception of the unsubstituted aromatic ring of apomorphine, the bulk of the molecular frameworks of the two compounds occupy approximately the same spatial area.

Subsequently, similar and independent conclusions by workers at the Eli Lilly laboratories led them to synthesize and test the pyrroleethylamine derivatives IV-VI (9). These all demonstrated ergoline-like dopaminergic activity, with the

IV   V

VI   VII

linear tricyclic compound VI (a "debenzergoline") most active. Most interesting was their finding that the pyrazole isostere VII possessed dopaminergic potency comparable to that of pergolide, yet lacked significant pharmacologic effects in other neurotransmitter systems. In sharp contrast was their subsequent finding that incorporation of a nitrogen into the 2-position of the ergolines to give 2-azaergolines abolished dopaminergic activity and led instead to weak dopamine receptor antagonist activity (10).

Complete removal of the pyrrole ring, as for example in the pergolide analogue VIII, led to greatly reduced dopaminergic activity (11). This marked attenuation in activity, as well as

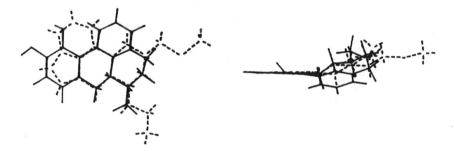

Figure 1. Computer-generated framework superposition (left) and edge-on view (right) of 6aR-apomorphine (——) and 5R-pergolide (– – –) as suggested by Nichols (8).

the general lack of dopamine-like action in nonhydroxylated octahydrobenzo[f]quinolines, further substantiates the role of the pyrrole ring in conferring dopaminergic activity on the ergolines.

Nevertheless, Cannon et al. have prepared several hydroxylated octahydrobenzo[f]quinolines which are very potent dopaminergics (12). These findings have generated considerable controversy regarding which portions of the ergolines are necessary for activity.

VIII   IX

The situation seems further confounded by the reports (13) of dopaminergic activity for the 4-substituted aminoethylindole DPAI, IX. However, the delayed onset of action reported by Cannon et al. (13) for this compound, as well as a weak in vitro action (14), lead to the possibility that DPAI may be metabolically activated by hydroxylation at the 6-position. This is a common transformation for indoles. Indeed, lergotrile is hydroxylated at the corresponding 13 position to yield a metabolite which is an order of magnitude more potent than lergotrile istelf (15).

Rigidification

Attempts to rigidify the dopamine molecule are exemplified by the synthesis of 2-amino-6,7-dihydroxy-1,2,3,4-tetrahydronaphthalene by Thrift in 1967 (16). Woodruff (17) suggested that this material, (6,7-ADTN) X, might be a dopamine agonist. Subsequent reports by Miller et al. (19) and Woodruff et al. (18) indeed demonstrated the potent dopamine like properties of X.

Cannon, in 1975, suggested that dopamine might bind to its receptors in two possible conformational extremes, which he designated the "alpha" and "beta" rotamers, XIa and XIb, respectively (20). In this view apomorphine would represent a structure containing a dopamine fragment "frozen" into the alpha rotameric form. By contrast, 6,7-ADTN contains a dopamine moiety constrained in the beta rotameric form. In both cases the side chain is nearly coplanar with the aromatic ring and is

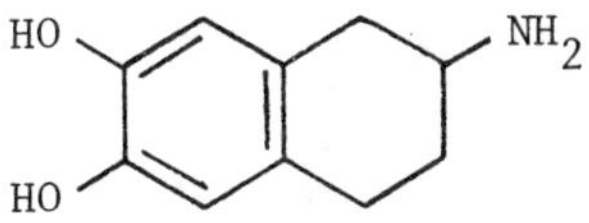

X

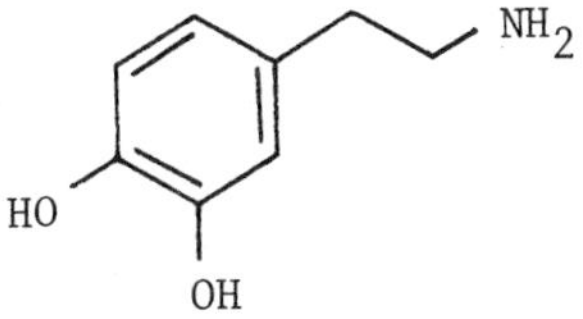

alpha rotamer XIa

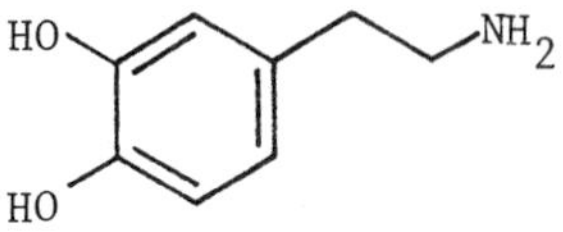

beta rotamer XIb

in a trans, extended, or antiperiplanar conformation. This provides a framework which places the nitrogen approximately into the aromatic ring plane, with the distance from the hydroxy which is "meta" to the side chain being 6.5–7.5 Å. Although examples will be discussed later where this does not apply, in all of them the basic amino group can be placed within about 1 Å of the aromatic ring plane.

The alpha and beta nomenclature has been convenient to classify structures which clearly contain a dopamine like fragment. Ignoring for the moment subtypes of receptors, it seems the data have generally supported the idea that the alpha rotameric form of dopamine is more important for emetic activity and activity at DA_2 receptors, while the beta rotameric form is more essential for action at the renal DA_1 receptor. Some important exceptions to this generalization will be presented later.

Costall et al. (21) have questioned the relevance of this alpha and beta nomenclature. These workers noted that many of the differences observed between dopamine congeners of the two types in eliciting peripheral and behavioral effects may be attributed to ability to penetrate into the CNS, differential distribution, and differing susceptibility to inactivation by COMT or MAO. The question of relative potency in comparing alpha rotameric types with beta rotameric types is heavily dependent on which type of bioassay is employed. This problem is compounded by the fact that a large number of novel dopamine agonist types have not been completely evaluated pharmacologically. Each laboratory seems to have a particular series of assays and it is seldom that a compound is examined in all relevant tests.

Several interesting rigid dopamine analogs have shown a lack of significant activity. Noteworthy among these are the trans-2-(3,4-dihydroxyphenyl)cyclopropyl and cyclobutyl amines XII and XIII (22,23). This lack of activity, and the loss of

activity seen in alpha methyl dopamine (24) led Ehrhardt to
suggest that the dopamine receptor cannot tolerate steric bulk
attached to the carbon adjacent to the amino group (25). This
idea is reinforced by the finding that 2-methyl-6,7-ADTN XIV is
inactive in the renal artery model (26).

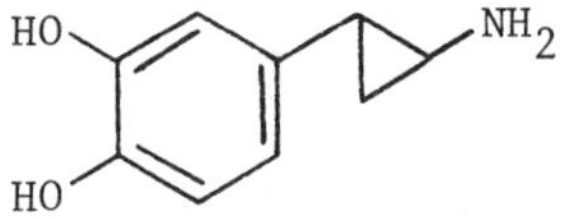

XII

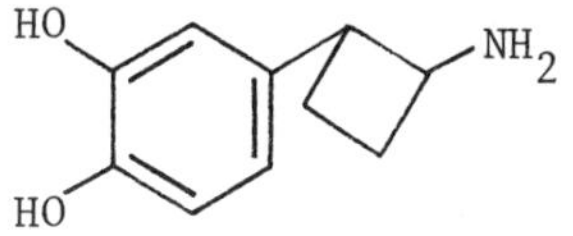

XIII

The process of rigidification has been extended, especially
by Cannon's group, to include a large series of cis and trans
fused octahydrobenzo[f] and octahydrobenzo[g]quinolines. Cannon
et al. (27) recently summarized this work. It is certainly
clear that incorporation of dopamine into a rigid framework can
alter both specificity and potency. However, suitable explanations for some of the observed activities are still lacking.
For example, the octahydrobenzo[g]quinoline XV (R=H or n-propyl)
is inactive in the renal DA$_1$ receptor (27). No one has so far
been able to explain why this is so, although a receptor model
is offered later is this paper which can accommodate this
finding.

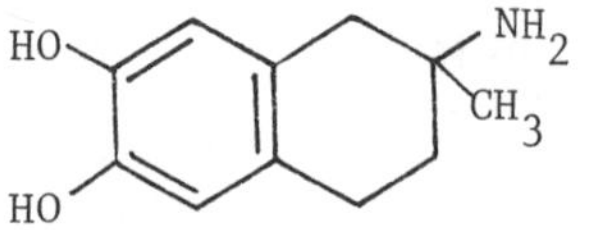

XIV

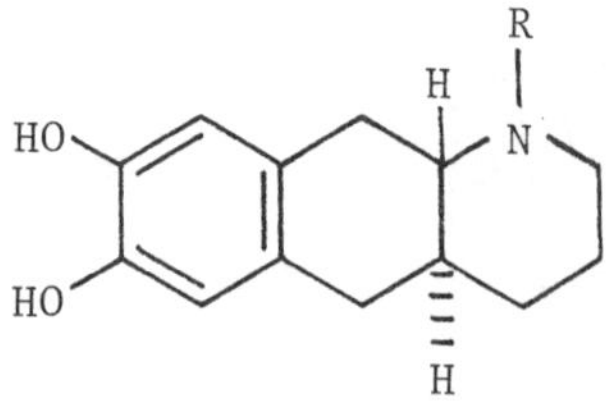

XV

Nitrogen Substitution

The effects of N-substituents on dopamine congeners are numerous and well documented. Generally, compounds active at DA_2 receptors retain or have enhanced activity as tertiary amines. By contrast, compounds active at DA_1 receptors generally are most active as primary, or sometimes secondary amines with one major exception: substitution of the nitrogen with di-n-propyl groups leads to compounds active at DA_1 receptors. This effect is so powerful that even in dopamine congeners which possess the so-called alpha rotameric moiety, not normally seen in DA_1 agonists, potent action is observed (28).

A single n-propyl is quite effective in restoring activity to many series of dopamine like compounds. Indeed, this unique alkyl group optimizes activity when attached to the amino function not only of dopamine-like structures, but also to the ergolines. We have dubbed this the "n-propyl phenomenon" and there seems no obvious explanation for its effect. It is clearly not a case of general hydrophobicity or passive diffusion, since activity through the series methyl-ethyl-propyl is not continuous. It seems to be a specific interaction with a hydrophobic site near the amine binding region, which perhaps has a unique geometry to accommodate a propyl.

Stereochemistry and Absolute Configuration

The importance of resolution and determination of absolute configuration cannot be overemphasized. There was, in this writer's opinion, little significant progress in developing useful receptor models prior to the determination of the absolute configurations for the active enantiomers of apomorphine, I, certain N-substituted 5-hydroxy-2-amino-1,2,3,4-tetrahydronaphthalenes, and of 6,7-ADTN (X). It is very common to see structures drawn in the literature with their chiral center shown as a particular absolute configuration, for example similar to that of apomorphine. Yet, in many of these cases there is no evidence as to which isomer is active. The reversed stereochemistry for the active enantiomers of apomorphine and 6,7-ADTN should serve as a warning not to assume that structures always correlate the way one might a priori expect.

The most useful model of the dopamine receptor(s?) has evolved largely from a concept proposed by McDermed et al. (29). This model was dependent on a knowledge of the absolute configurations for the active enantiomers of apomorphine and 6,7-ADTN. For both compounds, the R enantiomers are active. Sequence priority rules assign both as having the R configuration, but in fact the stereochemistries of the two compounds are reversed at their chiral carbon atoms. Thus, McDermed suggested that the two different rotameric types bind to the receptor as shown below. Using the nomenclature suggested by Rose et al.

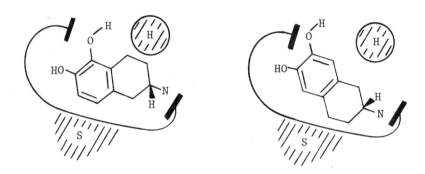

(30) for naming ring faces, the S-(-) enantiomer of 5- or 5,6-substituted 2-aminotetralins would interact with the "alpha" face, while R-(+)-6,7-ADTN would interact with its "beta" face. Priority rules which apply in this nomenclature designate the beta face of apomorphine as the one proposed to interact with the receptor, although this corresponds to the alpha face of M-7 (II).

Grol and Rollema had earlier proposed that there was a steric boundary on the receptor, which prevented interaction with compounds having bulk in this region (31). McDermed incorporated this feature, designated as "S" in the diagram above, into his dopamine receptor model. There are a large number of inactive compounds whose lack of activity can be explained by this feature. This region apparently lies adjacent to the C1-C8 edge of 5,6-dihydroxy substituted tetralins or the C4-C5 edge of 6,7-ADTN. Consistent with this feature, McDermed recently reported that 8-propyl-6,7-ADTN retains good activity, while the 5-propyl isomer is nearly inert (32). A modification of this model has been applied to the D-2 receptor by Seeman (33).

Hydrophobic features of the receptor

The retention of activity by 8-propyl-6,7-ADTN noted above leads to a consideration of a possible hydrophobic site or region of bulk tolerance on the receptor at that approximate site. A compound with the most obvious requirement for this is apomorphine, which would place its unsubstituted phenyl ring into this area. Ehrhardt has suggested (34) that this interaction may be very important. However 6,7-ADTN lacks this feature and it may be more important at D-2 or DA$_2$ receptors. The general location of this site is shown above as "H".

It seems possible that the presence of a hydrophobic group, such as a phenyl, on the agonist at this location may be one feature which differentiates between an action at DA$_1$ and DA$_2$ receptors. Two prime examples to consider which seem paradoxical in this context are the 1-phenyl-3-benzazepine XVI (SKF 38393) (35) and the 4-phenyl-1,2,3,4-tetrahydroisoquinolines XVII (36). Both of these possess a phenyl ring which could project into the postulated hydrophobic region of the receptor. However, an analysis of the probable conformations

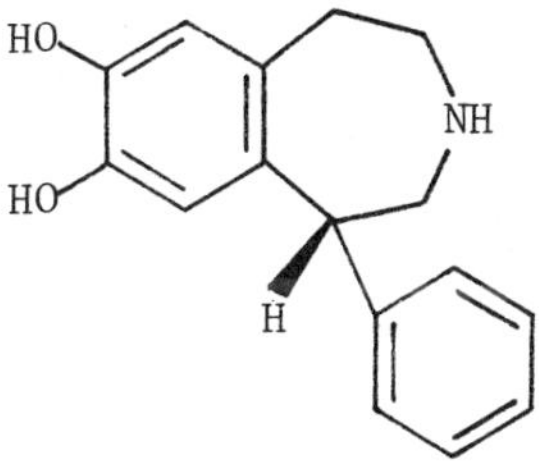

XVI 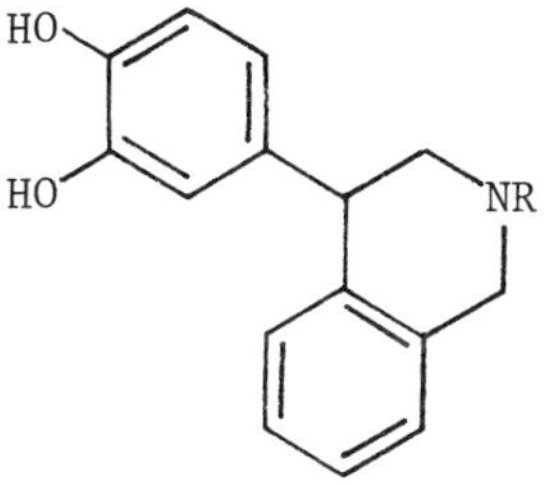   XVII

for each reveals that the non-hydroxylated aromatic ring is probably twisted out of the plane of the hydroxylated ring. Although the crystallographic structure for the benzazepine does not seem too helpful in understanding possible receptor interactions (37), the azepine ring has considerable conformational flexibility. If one assumes that a beta rotameric-like conformation for XVI and XVII is the one which interacts with DA$_1$ receptors, both congeners can be flexed to approximate this. Furthermore, the nonhydroxylated rings are forced away from the postulated receptor surface when these conformations are adopted. A framework stereopair superposition of XVI and XVII is illustrated in Figure 2, with the catechol oxygen and nitrogen atoms represented as spheres. Structure XVI is shown as a dashed structure and XVII is drawn in solid lines. The nitrogen atoms are not completely superimposed in these illustrations, but Dreiding models indicate that flexing the reduced rings of both systems allows relatively good correspondence. SKF 38393 is shown as the more active R enantiomer. It should be noted that the phenylisoquinoline, while also shown as the S enantiomer, has not as yet been resolved. Perhaps one can speculate that DA$_1$ agonists may possess bulk in this region, but it must not protrude into the plane of the catechol ring. Further, it seems possible that the presence of this phenyl ring

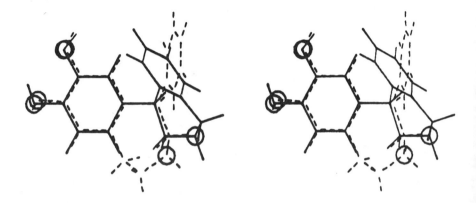

Figure 2.   Stereopair framework superposition of the R enantiomers of Structures XVI and XVII, with the oxygen and nitrogen atoms represented as circles. No attempt was made to simultaneously flex the reduced rings in order to achieve an optimum fit between the nitrogen atoms. The 1-phenyl group of XVI can also be rotated to achieve a better correspondence to the aromatic ring in XVII.

in SKF 38393 and related compounds may prevent interaction with adrenergic receptors and lead to the observed dopamine receptor selectivity (35).

For DA_2 receptor agonists, activity may be enhanced by the presence of a hydrophobic binding group at this site. For example, in N-(n-propyl)-N-2-phenethyl substituted compounds, the phenyl ring of the 2-phenethyl substituent could interact with this receptor feature.

Importance of nitrogen electron pair orientation

A comparison of XVI and XVII as DA_1 agonists has led to an additional interesting and perhaps important finding. The addition of an N-n-propyl group to either XVI or XVII does not lead to compounds with good activity. Indeed, an n-propyl effectively abolishes dopaminergic activity in XVII. This type of effect is very unusual among dopamine agonists. Since an N-ethyl also attenuated activity, it appeared to be a continuous function, rather than a sharp cutoff which might be due to steric bulk or length. An obvious explanation is the possibility that the N-alkyl can affect the conformation of the reduced ring.   One would expect the n-propyl to assume a pseudoequatorial conformation. Thus, in Figure 2, an n-propyl attached to either XVI or XVII would exist in a most favorable conformation when directed toward the lower right of the Figure. What is the significance of this observation? If one assumes that the real importance of the nitrogen atom lies in its unshared electron pair, whether free or protonated, then the orientation of this unshared pair will be important for optimum receptor interaction. Suppose, by way of speculation, that for the DA_1 receptor the electron pair must be able to reside in a pseudoequatorial orientation, and directed away from the "meta" hydroxy. In Figure 2 this would be the direction illustrated by the N-H vector shown as a solid line in XVII and the short segment of dashed line in XVI. This is also consistent with arguments for electron pair directionality proposed by Goldberg et al. (38). Note that for a secondary amine, or for a small alkyl, the nitrogen would be able to undergo inversion. A larger N-alkyl group will prefer to occupy the pseudoequatorial position and the unshared electron pair will be forced to reorient to a pseudoaxial position so that the direction vector for the lone pair will be approximately perpendicular to the plane of the hydroxylated aromatic ring.

This could explain the lack of activity observed by Cannon et al. for the octahydrobenzo[g]quinoline XV (27). If no N-substituent is attached, the electron pair could orient pseudoequatorially, but directed nearly 180° from the orientation attainable by XVI or XVII. Addition of an n-propyl would force the electrons into the pseudoaxial conformation.

With one major exception (38) it seems that little consideration has been given to the potential importance of the nitrogen electron pair directionality in dopamine agonists. However, based on this preliminary observation the following hypotheses are offered; when compounds with the so-called alpha rotameric configuration bind to the D-2 or DA_2 receptor, they may possess steric bulk in the plane of the hydroxylated ring, and which is located in a region corresponding to the non-hydroxylated ring of apomorphine. The orientation of the electron pair at this receptor is not critical. The N electrons may be directed normal to the plane of the agonist molecule. Further, it is possible that the three elements of: 1. a coplanar hydrophobic accessary group, 2. electron pair orientation, and 3. distance between the O_{meta}-N (ca. 6.7 Å) may be interrelated. That is, the presence of the hydrophobic group and/or the N to O distance may induce a conformation in the receptor which forces the receptor binding site for the nitrogen to approach from a particular direction. For compounds which contain the beta rotameric element, steric bulk in a region corresponding to the unsubstituted ring of apomorphine is not tolerated unless it can be projected above the plane of the agonist molecule. To be active at the DA_1 receptor, the nitrogen electrons must be oriented in an approximately pseudo-equatorial conformation, directed away from the "meta" hydroxy. Again, the O_{meta} to N distance (7.5 Å) may be interrelated to the requirement for a particular electron pair orientation of the nitrogen, as discussed above.

It should be emphasized again that the electron pair may be protonated at the receptor (i.e. formally an N-H bond). However this does not negate the orientation argument. If a proton is transferred from the agonist to a nucleophilic site on the receptor, the N-H bond orientation will be critical.

Unfortunately, there are several qualifications to these requirements, particularly as they apply to a discussion of DA_1 receptors. This receptor appears to have a requirement for two catechol hydroxys. There is no explanation for this apparent absolute requirement. The DA_2 receptor is less specific, requiring only one hydroxy. In the case of the ergolines even one hydroxy is not essential.

The second, and major, problem with this hypothesis is the fact that N-substitution with propyl groups can ruin concepts of orderly structure-activity relationships. The most glaring example of this is the activity of N,N-di-(n-propyl)-5,6-ADTN in the DA_1 receptor of the renal artery (28). In this case the electron pair must be oriented normal to the plane of the molecule and should not, by the above reasoning, be active. Empirical calculations reveal that the propyl groups essentially "lock" the nitrogen so that it cannot rotate about the C(2)-N bond. Thus, the presence of di-n-propyl groups constitutes a unique feature which must somehow distort the receptor so that electron orientation has little importance. Indeed, even non-hydroxylated derivatives possess activity when they are N,N-di-n-propyl substituted (39).

The ergolines generally conform to this receptor model. As illustrated in Figure 1, there can be a relatively good super-position of the apomorphine and ergoline structures. Camerman and Camerman (40) and also Cannon (12) have argued that it is possible that ring A of the ergolines corresponds to the dihydroxy substituted ring of the aporphines. This gives a superposition of the apomorphine and ergoline molecules which requires a conformational flip of the ergoline D ring to achieve any reasonable correspondence. However, Camerman and Camerman recently suggested (41) an alternative view consistent with this author's earlier proposal (8) and which seems to give a good fit to the receptor model. This correspondence is illustrated in Figure 3, by the superposition of 6a(R)-apomorphine with 5(R)-pergolide. The basic nitrogen atoms for the two structures are coincident, but the N_1H of pergolide has been placed approxi-mately at the location of the more important 11-OH ("meta" hydroxy) of apomorphine. Note that the A ring of the ergolines is displaced in the direction of the postulated hydrophobic region at this site on the receptor, as discussed earlier. This model of the receptor would also tolerate the 2-bromo or 2-chloro atoms of bromocriptine or lergotrile.

A consideration of this latter view suggests that the pyrrole NH of the ergolines is essential for their dopaminergic activity, as supported by the synthesis and testing of the partial ergoline structures noted earlier. However, how can one reconcile the fact that 13-hydroxy-lergotrile is an order of magnitude more potent as a dopamine agonist in vitro than its non-hydroxylated parent (15)? Also, although the non-hydroxylated octahydrobenzo[f]quinolines are inert as dopamine agonists, certain ring hydroxylated derivatives are very potent dopaminergics (12). It would seem that a phenolic hydroxy may be more effective in activating the receptor than the weakly acidic pyrrole NH. Hydroxylation of the ergolines in the 13 position, which corresponds to the 6 position of indole, leads to compounds which may orient on the receptor so as to align this hydroxy as the acidic function, rather than the pyrrole NH. This would correspond to the stereopair superposition shown in Figure 4 for the active isomers of apomorphine and 13-hydroxy-pergolide. Inspection of this view would suggest that the receptor may be able to accommodate the bulk of the pyrrole ring of the ergolines in a position which would correspond to the C(8)–C(9) edge of apomorphine. Metabolic hydroxylation may be an activating mechanism for dopaminergic ergolines which should be considered in the overall context of in vivo activity . It would seem prudent therefore, to tread lightly when attempting to "dissect out" the pharmacophore of the ergolines.

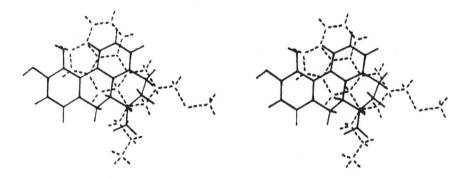

Figure 3. Stereopair framework superposition of the R enantiomers of apomorphine (———) and pergolide (– – –) by using the recent proposal of Camerman and Camerman (41).

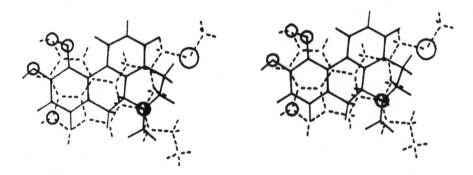

Figure 4. Stereopair framework superposition of the R enantiomers of apomorphine (———) and 13-hydroxypergolide (– – –) so as to correlate the ergoline 13-hydroxy with the 11-hydroxy of apomorphine. Oxygen, sulfur, and nitrogen atoms are represented as circles.

Conclusions

Hopefully these attempts to point out some possible additional features of dopamine receptors will be useful to the design of more specific agonists. This field suffers from the same problems as nearly all other areas of drug design, namely lack of knowledge about the receptor(s). On the other hand, with the level of effort which recently has been concentrated in this area, it seems likely that significant additional information will be gained in the next few years. It may be possible to develop very good empirical models of the dopamine receptor, at least for the purpose of designing dopamine agonists or antagonists. This author would like to emphasize the importance of determining absolute configuration for active dopamine agonists. Some of the speculation concerning superpositions of ergolines and aporphines should emphasize to medicinal chemists that they need to see these molecules less through the eyes of the printer, where benzene appears as a hexagon with formal Kekule bonds, and more as the picture the receptor "sees", of electrostatic fields and electron densities.

Literature Cited

1. Gund, P. Trends in Pharmacol. Sci. 1982, 3, 56–59.
2. Pinder, R. M.; Buxton, D. A.; Green, D. M. J. Pharm. Pharmacol. 1971, 23, 995–996.
3. Canon, J. G.; Kim, J. C.; Aleem, M. A.; Long, J. P. J. Med. Chem. 1972, 15, 348–350.
4. Long, J. P.; Heinz, S.; Cannon, J. G.; Kim, J. J. Pharmacol. Exp. Ther. 1975, 192, 336–342.
5. Cannon, J. G. Acta Pharm. Suecica 1982, in press.
6. Pieri, L.; Pieri, M.; Haefely, W. Nature 1974, 252, 586–588.
7. Von Hungen, K.; Roberts, S.; Hill, D. F. Nature 1974, 252, 588–589.
8. Nichols, D. E. J. Theor. Biol. 1976, 59, 167–177.
9. Bach, N. J.; Kornfeld, E. C.; Jones, N. D.; Chaney, M. O.; Dorman, D. E.; Paschal, J. W.; Clemens, J. A.; Smalstig, E. B. J. Med. Chem. 1980, 23, 481–491.
10. Bach, N. J.; Kornfeld, E. C.; Clemens, J. A.; Smalstig, E. B.; Frederickson, R. C. A. J. Med. Chem. 1980, 23, 492–494.
11. Bach, N. J.; Kornfeld, E. C.; Clemens, J. A.; Smalstig, E. B. J. Med. Chem. 1980, 23, 812–814.
12. Cannon, J. G. "Advances in Biosciences"; Imbs, J.-L.; Schwartz, J., Eds.; Pergamon, New York, 1979, Vol. 20, p. 87.
13. Cannon, J. G.; Demopoulos, B. J.; Long, J. P.; Flynn, J. R.; Sharabi, F. M. J. Med. Chem. 1981, 24, 238–240.
14. Clemens, J. A.; Kornfeld, E. C.; Phebus, L. A.; Shaar, C. J.; Smalstig, E. B.; Cassady, J. M.; Nichols, D. E.; Floss,

H. G.; Kelly, E. Proc. Symp. on Chemical Regulation of Biol. Mech., Cambridge, U.K., 1981.
15. Parli, C. J.; Schmidt, B.; Shaar, C. J. Biochem. Pharmacol. 1978, 27, 1405-1408.
16. Thrift, R. I. J. Chem. Soc. 1967, 288-293.
17. Woodruff, G. N. Comp. Gen. Pharmacol. 1971, 2, 439-455.
18. Woodruff, G. N.; Elkhawad, A. O.; Pinder, R. M. Eur. J. Pharmacol. 1974, 25, 80.
19. Miller, R.; Horn, A.; Iversen, L.; Pinder, R. M. Nature, 1974, 250, 238-241.
20. Cannon, J. G. Adv. in Neurology 1975, Vol. 9, p. 177.
21. Costall, B.; Lim, S. K.; Naylor, R. J.; Cannon, J. G. J. Pharm. Pharmacol. 1982, 34, 246-254.
22. Erhardt, P. W.; Gorczynski, R. J.; Anderson, W. G. J. Med. Chem. 1979, 22, 907-911.
23. Miller, D. D. Fed. Proc. 1978, 37, 2392-2395.
24. Wepierre, J.; Doreau, C.; Papin, A.; Paultre, C.; Cohen, Y. Arch. Int. Pharmacodyn. Ther. 1973, 206, 135-149.
25. Erhardt, P. W. J. Pharm. Sci. 1980, 69, 1059-1061.
26. Nichols, D. E.; Kohli, J.D., unpublished results.
27. Cannon, J. G.; Long, J. P.; Bhatnagar, R. J. Med. Chem. 1981, 24, 1113-1118.
28. Kohli, J. D.; Goldberg, L. I.; McDermed, J. D. Eur. J. Pharmacol. 1982, in press.
29. McDermed, J. D.; Freeman, H. S.; Ferris, R. M. "Catecholamines: Basic and Clinical Frontiers"; Usdin, E.; Kopin, I.J.; Barchas, J.D., Eds.; Pergamon: New York, 1979, p. 568.
30. Rose, I. A.; Hanson, K. D.; Wilkinson, K. D.; Wimmer, M. J. Proc. Natl. Acad. Sci., USA 1980, 77, 2439-2441.
31. Grol, C. J.; Rollema, H. J. Pharm. Phrmacol. 1977, 29, 153-156.
32. McDermed, J. Acta Pharm. Suecica 1982, in press.
33. Seeman, P. Pharmacol. Rev. 1980, 32, 229-313.
34. Ehrhardt, P.W. Acta Pharm. Suecica 1982, in press.
35. Setler, P. E.; Sarau, H. M.; Zirkle, C. L.; Saunders, H. L. Eur. J. Pharmacol. 1978, 50, 419-430.
36. Jacob, J. N.; Nichols, D. E.; Kohli, J. D.; Glock, D. J. Med. Chem. 1981, 24, 1013-1015.
37. Kaiser, C.; Dandridge, P. A.; Garvey, E.; Hahn, R. A.; Sarau, H. M.; Setler, P. E.; Bass, L. S.; Clardy, J. J. Med. Chem. 1982, 25, 697-703.
38. Goldberg, L. I.; Kohli, J. D.; Kotake, A. N.; Volkman, P. H. Fed. Proc. 1978, 37, 2396-2402.
39. Rusterholz, D. B.; Long, J. P.; Flynn, J. R.; Cannon, J. G.; Lee, T.; Pease, J. P.; Clemens, J. A.; Wong, D. T.; Bymaster, F. P. Eur. J. Pharmacol. 1979, 52, 73-82.
40. Camerman, N.; Chan, L. Y. Y.; Camerman, A. Mol. Pharmacol. 1979, 16, 729-736.
41. Camerman, N.; Camerman, A. Mol. Pharmacol. 1981, 19, 517-519.

RECEIVED February 16, 1983

Commentary: The Development of Novel Dopamine Agonists

ALAN S. HORN

University of Groningen, Department of Pharmacy, Groningen, The Netherlands

The author of the previous article has highlighted some of
the salient features that have emerged over recent years with regard to structure activity relationships of dopamine receptor agonists and their application to the design of new agents.

In his introduction he mentions in passing the problem of DA
receptor classification and the difficulties it has produced for
drug designers. Thus he is of the opinion that it "represents a
serious obstacle to design of selective dopamine agonists" and
that "The situation in the central nervous system, at best, is
needlessly complex at the present time". To a certain extent he is
correct, the situation is somewhat confused and as Beart has humo-
rously pointed out (1), like rabbits, some researchers just don't
know when to stop! However, I feel that the author's position is
a little too negative. For example, the autoreceptor concept of
Carlsson (2) has led to the development of a selective agonist, 3-
PPP, for this type of DA receptor (3). The D-1 and D-2 receptor
classification of Kebabian and Calne (4) has also led to the discovery of specific agonists such as SKF 38393 (D-1) and LY 141865
(D-2) (5, 6). The significance of his remark about the situation
in the CNS being "needlessly complex" is not entirely clear. The
apparent complexity may be intrinsic to the system and thus necessarily rather than "needlessly" complex or it may be an artifact
of the radioligand binding method.

Although the various DA receptor classifications have been
useful in the development of new agonists there have been some
"false positives". Thus in spite of the fact that DPI has been
claimed by Cools and van Rossum (7) to be a DA_i receptor agonist
it has been found to be inactive in so many dopaminergic test systems that it should no longer be considered as a normal DA agonist
(8, 9). Similarly the claim (10) that TL-99 is a selective presynaptic DA agonist is doubtful because it has been shown that the
hypomotility it causes in rats is partly due to its potent α_2-
adrenergic activity (11). In addition it has been found to have
postsynaptic activity (12).

In his section on structural dissection Nichols comments

0097–6156/83/0224–0219$06.00/0
© 1983 American Chemical Society

on the elegant work of the Eli Lilly group with pyrrole and pyra-
zole-ethylamine analogues as DA agonists (13). This is certainly
one of the most interesting and possibly important findings in
this area in recent years. In a similar manner the recent demonstration that the nitrogen atom in DA can be replaced by a positively charged sulfur atom is also of considerable significance
(14).

The concept of alpha and beta rotameric forms of DA is
covered in the section on rigidification. A summary of some recent
findings in this area together with details of the X-ray structure
of 6,7-ADTN has also been published by the present author (15). In
connection with the above topic our group has published quantitative details of the penetration and distribution of DA agonists
into the CNS as well as the effect of COMT inactivation (16, 17,
18). The author mentions the work of Cannon who has carried out
some of the best studies on the effects of rigidification of DA
analogues. The latter compounds are certainly much better examples
of this concept than compounds XII and XIII which are not really
"rigid" but merely analogues with restricted conformational free-
dom.

One of the most intriguing aspects of DA structure-activity
relationships is the so-called "N-propyl effect" and this is discussed in the section on N-substitution. It is well known that in
many diverse series of DA analogues those having an N-propyl substitution often exhibit maximum activity. As Nichols points out
this must be due to a specific interaction with a hydrophobic site
near the N atom. It is also of interest that in a later section he
mentions some remarkable exceptions to this general rule.

The following section is devoted to stressing the importance
of a knowledge of the stereochemistry and absolute configuration
of DA agonists. One can only fully agree with Nichols when he
says that such information is invaluable in increasing our understanding of the subtleties of DA receptor interactions.

The hydrophobic features of the DA receptor is the following
topic and this is an area that can be very usefully exploited by
the medicinal chemist in his search for new analogues via lead
optimalization. Its importance has been pointed out by various
authors (19, 20).

In his final section Nichols discusses the concept of the importance of the nitrogen lone pair orientation. Although he cor-
rectly points out that this idea was first suggested for DA analogues by Goldberg et al. (21) it has, in fact, been raised some
time ago in connection with a similar problem in another area,
i.e., the binding of opiates to their receptor sites (22, 23).
From studies with quaternary compounds it has been concluded,
however, that it is probably the positively charged form of the
opiates that binds to the receptor(s) (24). Bearing in mind the
interesting results with the previously mentioned sulfonium analogue of DA (14) the significance of the nitrogen lone-pair orientation theory is open to question.

Literature Cited

1. Beart, P.M. Trends Pharm. Sci. 1982, 3, 100-102.
2. Carlsson, A. "Pre- and Postsynaptic Receptors", Usdin, E. and
   Bunney, W.E., Eds., Marcel Dekker, New York 1975, 49.
3. Hjorth, S.; Carlsson, A.; Wikström, H.; Lindberg, P.; Sanchez,
   D.; Hacksell, U; Arvidsson, L.-E.; Svensson, U.; Nilsson,
   J.L.G. Life Sci. 1981, 28, 1225-1232.
4. Kebabian, J.W.; Calne, D.B. Nature (London) 1979, 277, 93-96.
5. Setler, P.E.; Sarau, H.M.; Zirkle, C.L.; Saunders, H.L. Eur.
   J. Pharmacol. 1978, 50, 419-430.
6. Tsuruta, K.; Freg, E.A.; Grewe, C.W.; Cote, T.E.; Eskay, R.L.;
   Kebabian, J.W. Nature (London) 1981, 292, 463-465.
7. Cools, A.R.; van Rossum, J.M. Psychopharmacologia 1976, 45,
   243-254.
8. Van Oene, J.C.; Houwing, H.A.; Horn, A.S. Eur. J. Pharmacol.
   1982, 81, 75-87.
9. Van Oene, J.C.; Houwing, H.A.; Horn, A.S. Eur. J. Pharmacol.
   1982, in press.
10. Goodale, D.B.; Rusterholz, D.B.; Long, J.P.; Flynn, J.R.;
    Walsh, B.; Cannon, J.G.; Lee, T. Science 1980, 210, 1141-1143.
11. Horn, A.S.; de Vries, J.; Dijkstra, D.; Mulder, A.H. Eur. J.
    Pharmacol. 1982, 83, 35-45
12. Martin, G.E.; Haubrich, D.R.; Williams, M. Eur. J. Pharmacol.
    1981, 76, 15-23.
13. Bach, N.J.; Kornfeld, E.C.; Jones, N.D.; Chaney, M.O.;
    Dorman, D.E.; Paschal, J.W.; Clemens, J.A.; Smalstig, E.B.J.
    J. Med. Chem. 1980, 23, 481-491.
14. Anderson, K.; Kuruvilla, A.; Uretsky, N.; Miller, D.D. J. Med.
    Chem. 1981, 24, 683-687.
15. Horn, A.S.; Rodgers, J.R. J. Pharm. Pharmacol. 1980, 32,
    521-524.
16. Westerink, B.H.C.; Dijkstra, D.; Feenstra, M.G.P.; Grol, C.J.;
    Horn, A.S.; Rollema, H.; Wirix, E. Eur. J. Pharmacol. 1980,
    61, 7-15.
17. Rollema, H.; Westerink, B.H.C.; Mulder, T.B.A.; Dijkstra, D.;
    Feenstra, M.G.P.; Horn, A.S. Eur. J. Pharmacol. 1980, 64,
    313-323.
18. Horn, A.S.; Dijkstra, D.; Mulder, T.B.A.; Rollema, H.;
    Westerink, B.H.C. Eur. J. Med. Chem. 1981, 16, 469-472.
19. Hacksell, U.; Svensson, U.; Wilsson, J.L.G.; Hjorth, S.;
    Carlsson, A.; Wirkström, H.; Lindberg, P.; Sanchez, D. J. Med.
    Chem. 1979, 22, 1469-1475.
20. Seeman, P. Pharmacol. Rev. 1980, 32, 229-313.
21. Goldberg, L.I.; Kohli, J.D.; Kotake, A.N.; Volkman, P.H. Fed.
    Am. Soc. exp. Biol. 1978, 37, 2396-2402.
22. Belleau, B.; Conway, T.; Ahmed, F.R.; Hardy, A.D. J. Med.
    Chem. 1974, 17, 907-908.
23. Belleau, B; Morgan, P. J. Med. Chem. 1974, 17, 908-909.
24. Opheim, K.E.; Cox, B.M. J. Med. Chem. 1976, 19, 857-858.

RECEIVED January 27, 1983

Stereoisomeric Probes of the Dopamine Receptor

CARL KAISER

Smith Kline & French Laboratories, Research and Development Division, Philadelphia, PA 19101

Enantiomers of various dopaminergic agents have been utilized to complement structure-activity studies to examine dopamine receptors. Optical antipodes are particularly useful in such studies as their identical physical properties assure that their pharmacological actions are likely due to receptor interactions, unless they are metabolized differently. In the present investigation enantiomers of 2,3,4,5-tetrahydro-7,8-dihydroxy-1-phenyl-1H-3-benzazepine (SK&F 38393), the 6-chloro-4'-hydroxy derivative (SK&F 82526) and its 3-allyl congener (SK&F 85174) were separated. Optical purity was determined by NMR and HPLC methods. Absolute stereochemistry was established by single-crystal x-ray diffractometric analysis and CD spectral methods. Dopamine-like activity was determined in several in vitro and in vivo tests. In certain cases a high degree of enantioselectivity was observed. In an attempt to rationalize the observed enantioselectivity, structure-activity relationships, conformational and configurational aspects of the benzazepines were compared to those of certain other dopamine receptor agonists and antagonists. Comparison of the benzazepines with various receptor models derived from other dopamine agonists/antagonists, permits rationalization of their enantioselectivity. The 1-phenyl substituent in the benzazepines apparently binds to an accessory lipophilic site; however, precise location of this region was not possible. It is suggested that the mode of binding of the benzazepines may differ somewhat from that of many other dopamine receptor antagonists and agonists. Conceivably, the different overall conformations of agents of this class may account for their mixed agonist-antagonist activities.

0097-6156/83/0224-0223$07.00/0
© 1983 American Chemical Society

The recognition that dopamine (DA) is an important neurotransmitter with receptors in both the central nervous system
and in the periphery has resulted in a widespread search for
substances that either mimic or block the action of this natural
neurotransmitter on its receptors. This has resulted in the
discovery of an enormous number of compounds of diverse structure that have DA receptor agonist and antagonist properties.
In many, but not all, instances it has been possible to relate
the structure of these compounds to that of DA, generally in a
trans (antiperiplanar, extended) conformation in which its salt
is found in the solid state (1).

Discovery of the DA-like activity of 2,3,4,5-tetrahydro-7,
8-dihydroxy-1-phenyl-1H-3-benzazepine (I, SK&F 38393) (2, 3) was
of interest for several reasons. Firstly, unlike many other DA
receptor modulators, conformational constraints imposed on this
molecule by virtue of the tetrahydroazepine ring system prevent
the embodied DA framework from attaining a fully extended conformation. In addition, this agent has a unique biological profile suggesting selectivity for the adenylate cyclase modulated
D-1 (4) subpopulation of DA receptors. Also, it acts as a mixed
agonist-antagonist in its ability to stimulate central DA-sensitive adenylate cyclase. These observations led to an extensive
examination of compounds related to I in an effort to identify
not only new selective DA receptor agonists (5), but also antagonists. As a consequence of this study 3-benzazepine relatives
of I with selectivity for peripheral DA receptors (6), as well
as some new antagonists (7), were discovered. Several unique
peripherally acting DA receptor agonists, namely II (SK&F 82526)
and III (SK&F 85174), have been the subject of detailed pharmacological examination (8-12). 6-Chloro-2,3,4,5-tetrahydro-7,8-
dihydroxy-1-(4-hydroxyphenyl)-1H-3-benzazepine (II) is presently being studied in the clinic (9).

I, R = X = H; Ar = C_6H_5

II, R = H; X = Cl; Ar = 4-HOC_6H_4

III, R = $CH_2CH=CH_2$; X = Cl;
       Ar = 4-HOC_6H_4

IV, R = X = Ar = H

Structure-activity relationship (SAR) studies in this series of tetrahydrobenzazepines (5) indicate that the 1-phenyl,
or substituted phenyl, group contributes significantly to the DA
receptor agonist properties of these compounds (13, 14). As the
1-phenyl substituent in I-III provides these molecules with an
asymmetric center, separation and study of the enantiomers was
of particular interest.

Enantiomers of both DA receptor agonists (e.g., 15-19) and

antagonists (e.g., 20-29) have uniformly demonstrated enantioselectivity. These findings have been employed to complement
SAR observations in probing the pharmacophores, their mode of
interaction with DA receptors and, in turn, to provide information relevant to the topography of the receptors. As enantio-
meric pairs have identical physical properties, if it is assumed
that the isomers are metabolized identically, biological differences (30-33) may reasonably be associated with receptor related events (34). These considerations, coupled with the pharmacological and chemical uniqueness of the tetrahydrobenzazepines I-III, suggested examination of their optical antipodes
for DA-like activity.

Chemistry

    Initially, the racemates of I-III were not easily separa-
ble by recrystallization of diastereoisomeric salts with opti-
cally active acids; however, methylated precursors (V a,b and VI
a,b) were easily resolved via their dibenzoyltartrates (35, 36).
Conversion of these precursors to the optical antipodes of I-III
was performed as illustrated in Scheme I (35, 36).
    Absolute stereochemistry of the enantiomers of I was es-
tablished by single-crystal x-ray diffractometric analysis of
(R)-I•HCl (37), as well as by similar analysis of (R)-Va•CH₃I
which was subsequently converted to (R)-I via a stereospecific
synthetic sequence (35). Computer generated perspective drawings (38) of two rotamers of the isomer [[α]$_D^{25}$ + 15.3° (c 1,
CH₃OH)] observed in the x-ray analysis are shown in Figure 1.
The location of the hydrogen in position 1 is included for clar-
ity. The absolute configurations of (R) and (S)-II and III were
assigned on the basis of comparison of their circular dichroism
spectra with those of (R) and (S)-I of known stereochemistry
(36). Subsequently, the absolute stereochemistry of (R)- II•HBr
was confirmed by single-crystal x-ray diffractometric analysis
(39). Optical purities of the enantiomers of I-II were determined by high performance liquid chromatographic analysis of
their amides with (-)-α-methoxy-α-trifluoromethylphenyl-acetic
acid. The optical purity of (R) and (S)-III was estimated by NMR
differentiation of methoxyl signals of its precursor VII in the
presence of a chiral shift reagent (36). In all instances enan-
tiomeric excesses were shown to be about 90% or greater (35, 36).

Pharmacology of (R) and (S)-I-III

    As indicated in Table I, I-III and their R and S optical an-
tipodes were studied for dopaminergic activity in four primary
tests: (1) stimulation of central DA-sensitive adenylate cyclase (2, 35), (2) displacement of [³H]-spiroperidol binding
to rat caudate homogenate (35, 40), (3) induction of rotations
in rats with unilateral lesions of the left substantia nigra (2,

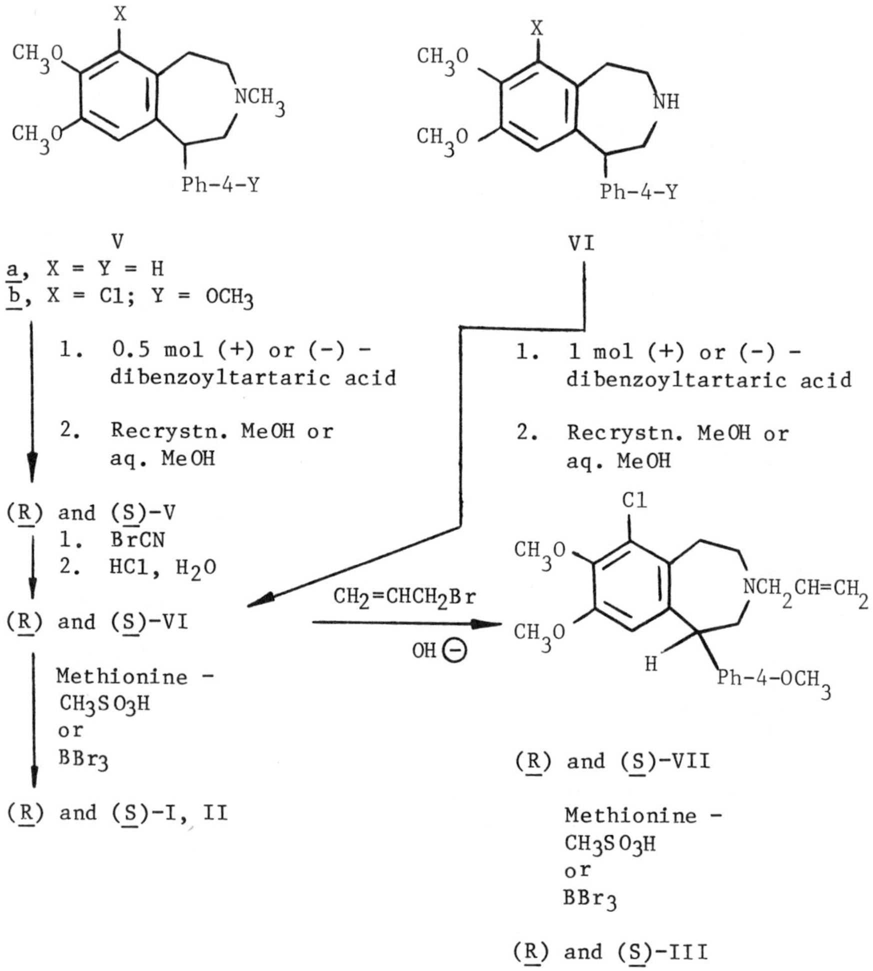

Structure V (left):
CH$_3$O, CH$_3$O, X, NCH$_3$, Ph-4-Y

V

a, X = Y = H
b, X = Cl; Y = OCH$_3$

Structure VI (right):
CH$_3$O, CH$_3$O, X, NH, Ph-4-Y

VI

(from V):
1. 0.5 mol (+) or (−) −
   dibenzoyltartaric acid

2. Recrystn. MeOH or
   aq. MeOH

(from VI):
1. 1 mol (+) or (−) −
   dibenzoyltartaric acid

2. Recrystn. MeOH or
   aq. MeOH

(R) and (S)-V
1. BrCN
2. HCl, H$_2$O

(R) and (S)-VI

CH$_2$=CHCH$_2$Br

OH ⊖

Structure VII:
CH$_3$O, CH$_3$O, Cl, H, NCH$_2$CH=CH$_2$, Ph-4-OCH$_3$

(R) and (S)-VII

Methionine −
CH$_3$SO$_3$H
or
BBr$_3$

(R) and (S)-I, II

Methionine −
CH$_3$SO$_3$H
or
BBr$_3$

(R) and (S)-III

Scheme I

35), and (4) reduction of renal vascular resistance (RVR) in dogs
(6). Potency in the adenylate cyclase test [a D-1 receptor response (4)] is expressed in terms of an EC$_{50}$, the molar concentration of compound that produces half maximal stimulation of
cyclic-AMP production. This maximum (% of maximum DA response)
and the concentration at which it is observed are also given.
Spiroperidol binding [a central DA response involving D-2 receptors (41)] effectiveness is presented as an IC$_{50}$, i.e.,
the concentration of compound causing 50% inhibition of specific spiroperidol binding. Rotation in rats with lesions in the
substantial nigra (a supersensitive DA receptor response) is
quantitated by a RD$_{500}$ value which is defined as the dose of

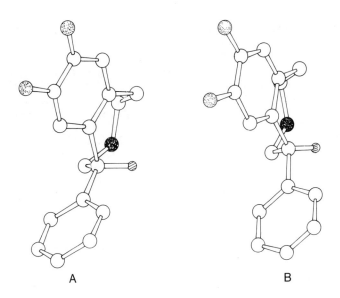

A   B

Figure 1.   Computer-generated perspective drawing of two rotamers (A,B). (See text for discussion) (37, 38). Key: ◯, carbon; ●, nitrogen; ⊕, oxygen; and ⊘, hydrogen.

compound calculated to produce 500 bodily rotations during a 2 hour test period. Renal vasodilator potency (presumably a peripheral DA receptor agonist action) is expressed as ED_{15} and ED_{20} values, the average maximum cumulative dose that decreases RVR by 15% and 20%, respectively. The maximum observed decrease in RVR and the cumulative dose producing this effect are also presented.

Structure-Activity Relationship Considerations

The data presented in Table I indicate that a properly oriented 1-phenyl substituent is important for DA receptor activation by 2,3,4,5-tetrahydro-7,8-dihydroxy-1H-3-benzazepines. Thus, in the test for stimulation of DA-sensitive adenylate cyclase the R isomers, as well as the racemates of I-III are considerably more potent than the parent IV which lacks a substituent in the 1 position. The EC_{50} of IV is 5.2 μM (35); i.e., it is two orders of magnitude less potent than I in this system (see Table I) (13). Likewise, IV was 5-fold less effective in causing contralateral rotations in rats with lesions in the substantia nigra when the compounds were administered directly into the ipsilateral striatum. The RD_{500} for I was 0.18 μg/rat as compared to 0.9 μg/rat for IV (14). It also appears that the 6-Cl substituent enhances binding at the receptor. Although complicated by differences in the substituents in positions 1 and 3, the observation that racemic II and III are more potent than I as stimulants of rat striatal adenylate cyclase is suggestive of potency enhancement by virtue of 6-Cl substitution. This suggestion is supported by the 8-fold enhancement of the potency of I in this test upon introduction of a 6-Cl substituent (6). The consequence of the 4'-OH substituent on receptor activation is less clear. Thus, 4'-hydroxylation of racemic I results in an approximately 10-fold decrease in potency in the adenylate cyclase assay and the observation (Table I) that (R,S)-II is only slightly more potent than I in this test is consistent with a negative contribution from the 4'-OH substituent. In contrast, racemic II is nearly two orders of magnitude more potent than I as a renal vasodilator in dogs. Although several rationalizations might be advanced for this finding, it is possible that the 4'-OH substituent increases the overall polarity of the molecule to hinder its accumulation in lipophilic sites with a resultant elevated concentration in the kidney (6). The peripheral selectivity of the 4'-OH derivatives II and III is evidenced by their effectiveness in the dog renal vasodilator test following iv administration whereas they are ineffective in the lesioned rat rotational model following ip dosing. These results suggest that these compounds are unable to penetrate the lipophilic "blood-brain barrier" to enter into the brain. This supposition is supported by the demonstration that II is quite potent (about one-fifth as potent as DA) in this test when it is given intracaudally (6).

Table I. Dopaminergic Activity of I-III and Enantiomers[a]

Compound	Adenylate cyclase ↑ c-AMP, EC_{50}, µM,[b,c] [max ↑,[d] % at concn, µM]	Spiroperidol binding IC_{50}, µM[b,c]	Rotation in lesioned rat, RD_{500},[b,c] mg/kg, ip or [rotations, mg/kg, ip]	Renal vasodilator, dogs, ED,[b] µg/kg, iv (max ↓ RVR, cum. dose)
(R,S)-I[e]	0.071 (0.052-0.106) [68,10]	34.43 (33.42-41.91)	0.7 (0.0-1.0)	$ED_{15}=31$[f] (19,110)
(R)-I[e]	0.032 (0.02-0.05) [66,10]	33.86 (28.56-40.14)	0.5 (0.33-0.71)	$ED_{15}=25$ (18,40)
(S)-I[e]	---- ---- [27,[g] 100]	197.4 (151.7-257.0)	[50+8,2.0][g]	[g] (9,550)
(R,S)-II	0.057 (0.031-0.096) [68,1.0]	1.03 (0.9-1.17)	[14+7,10][h]	$ED_{15}=0.3$ $ED_{20}=0.8$ (50,5300)
(R)-II	0.037 (0.022-0.064) [73,10]	0.54 (0.48-0.60)	[i]	$ED_{15}=0.31$ $ED_{20}=0.42$ (48,1500)
(S)-II	1.47 (0.79-2.7) [44,10]	18.2 (16.0-21.0)	[i]	$ED_{15}=0.94$ $ED_{20}=6.67$ (38,530)
(R,S)-III	0.015 (0.01-0.021) [85,1.0]	0.152 (0.132-0.176)	[647+84,10]	$ED_{20}=21.82$ (29,170)
(R)-III	0.0088 (0.005-0.013) [80,1.0]	0.106 (0.086-0.129)	[i]	$ED_{20}=0.75$ (51,20)
(S)-III	[18,[g] 10] ----	15.3 (13.3-17.5)	[i]	$ED_{15}=4.4$[f] (19,20)
DA[e]	3.5 [100,45] ----	5.34 (4.18-6.82)	[j]	$ED_{15}=2.7$ $ED_{20}=5.0$ (43,100)

Continued on next page.

Footnotes to Table I.

<u>a</u> All compounds were tested as CH_3SO_3H salts unless otherwise noted. <u>b</u> See text for references to test systems, EC, ED, RD and IC values. <u>c</u> 95% confidence limits in parentheses. <u>d</u> Percent of cyclic-AMP induced by 5 x 10^{-5} M DA. <u>e</u> HCl salt. <u>f</u> Insufficiently potent for determination of ED_{20} value. <u>g</u> Insufficiently potent for determination of EC, RD or ED values. <u>h</u> RD_{500} = 0.5 (0.4-0.8) µg/rat, i.c. <u>i</u> Not tested. <u>j</u> Value not available; inability of DA to enter CNS after ip administration requires use of DOPA and DOPA decarboxylase inhibitor (<u>36</u>); RD_{500} = 0.10 (0.08 - 1.3) µg/rat, i.c.

Introduction of an allyl group into position 3 of I and II apparently has little effect, or slightly increases, potency as stimulants of DA-sensitive adenylate cyclase. I and its N-allyl derivative (EC_{50} ca. 0.065 µM) are about equipotent in this test whereas racemic III is almost 4-fold more potent than racemic II. N-Allyl substitution of II, however, seems to modify the overall DA receptor agonist profile of the molecule. Comparison of the data for II and III in the adenylate cyclase and spiroperidol tests shows that II is about 18 times more potent in the adenylate cyclase (D-1) test than in displacing bound spiroperidol (D-2) (<u>41</u>) whereas III is only 10 times more effective. This may suggest that N-allylation results in modification of the affinity of the agents for subtypes of DA receptors. Indeed, secondary pharmacological studies indicate that the mechanism of action of III includes a presynaptic dopaminergic component (<u>10</u>). The possibility that II and III may have different effects on various receptors may also account for the differences in potency and overall cardiovascular profiles (<u>39</u>, <u>42</u>) of these two compounds in the dog renal vasodilator test.

In summary, the results of SAR studies suggest that substitution of positions 3, 6 and 4' influence the potency, selectivity and even the mechanism of action of 2,3,4,5-tetrahydro-7,8-dihydroxy-1H-3-benzazepines as DA receptor agonists.

<u>Configurational Considerations</u>. In general, the data tabulated in Table I indicate enantioselectivity for the benzazepines I-III. DA receptor agonist activity resides mainly in the R isomers; however, the R:S potency ratios vary considerably depending upon the particular test system. In some instances the S isomers are quite effective. This is particularly notable for II and III in the test for renal vasodilator activity in dogs. Comparison of the ED_{15} and ED_{20} values for (R) and (S)-II provides potency ratios of 3 and 15.9, respectively, and both isomers are somewhat more potent than I. Likewise, both (R) and (S)-III have a high degree of renal vasodilator activity in

anesthetized dogs. The reason for decreased enantioselectivity
of II and III in this in vivo test is not known. Perhaps it is
reflective of an altered effect of the isomers on different or
multiple receptors. This is consistent with the observation
that the overall cardiovascular profiles of the enantiomers of
II (42) and III (39) are significantly different. Another suggestion that the overall mechanisms of action of the enantiomers
of III may differ derives from the observation that racemic III
is 30-fold less potent than its R enantiomer. Perhaps (S)-III
interferes with the renal vasodilator efficacy of (R)-III. Apparently, however, this is not the case for (S)-II. (R)-II is
about equipotent with the racemate. The comparatively weak potency of (S)-III relative to that of (S)-II in the adenylate cyclase and dog renal vasodilator tests may suggest that the D-1
receptor may be able to accommodate both antipodes of II, although the S fits less readily than the R isomer. The relative
ineffectiveness of (S)-III may indicate that the N-allyl group
of this antipode interferes with D-1 receptor interaction. This
observation should be considered in models for D-1 receptors.
It appears of no significance in the spiroperidol displacement
test where (S)-II and (S)-III are approximately equipotent.
Thus the N-allyl group may be accommodated in D-2 receptors
(36), which may tolerate greater bulk in the vicinity of the
basic N.

   Conformational Considerations. A great deal of attention
has been directed toward the mode of interaction of DA with its
receptors. As a consequence of the multitude of conformations
that DA can attain because of its flexible aminoethyl chain, many
studies have been aimed toward more rigid, conformationally restricted analogs (e.g., 43-48). These studies have led to the
general conclusion that a nearly fully extended trans (antiperiplanar) conformation of the DA, as illustrated in the Newman
projection VIII, is required for both agonists (15) and antagonists (49) of the receptor. It has been suggested that DA may
interact with its receptors in two rotameric extremes, the so-
called α and β forms (43, 44). Several dihydroxyoctahydro-
benzo[f]quinolines are noteworthy because they rigidly fix the
catecholic system and the basic N, two groups generally acknow-
ledged to be critical for receptor interaction, in a steric relationship simulating the α- and β-rotameric forms. Thus, 7,
8-dihydroxyoctahydrobenzo[f]quinoline (IX) (45) closely ap-
proximates the α conformer of DA and the 8,9-dihydroxy isomer
(46) simulates the β-form. The trans isomers are very rigid
planar molecules whereas the cis isomers are capable of flip
(trans and gauche) conformations. Only the trans isomer of IX
causes potent central and peripheral dopaminergic actions (45).
The trans isomer of X, however, produces only potent peripheral,
but not central, DA-like effects (46). The distance between the
OH group "meta" to the ethylamine chain and the basic N is 6.4 Å

in IX (as it is in apomorphine). In X this distance is approximately 7.3 Å. It has been suggested this spatial difference may
explain the variations in reactivity of the compounds with DA
receptor subtypes (46). Several studies with somewhat more
flexible aminotetralins approximating the α and β rotameric
forms of DA have suggested that these conformations may account
for differences in activity in particular pharmacological models
(47, 50, 51). Despite these observations, derivation of a general conclusion has not been possible because classes of compounds simulating both the α- and β- conformers of DA clearly contain potent DA receptor agonists (47, 52). A recent comprehensive study of the behavioral effects of many rigid DA analogs suggests that definition of the preferred rotameric conformation for dopaminergic activity may be an "illusory quest" (53).
A final conformational point to be addressed is the relationship
of the catechol ring to the ethylamine chain; i.e., whether the
ring is perpendicular to or planar with the chain. The potential energy difference between these rotameric forms is small.
Molecular orbital studies and crystallographic analysis indicate
that the perpendicular form is preferred (54, 55). The activity
of the rigid analogs of DA, however, suggest that a coplanar arrangement is important.

VIII   IX   X

The conformation of I-III is unlike that of most other DA
receptor agonists. As a result of the conformational restraints
imposed by the tetrahydroazepine ring the DA skeleton within
these compounds is unable to attain the extended form of DA
(VIII). Nevertheless, the tetrahydroazepine ring of I-III does
permit considerable conformational flexibility; the DA framework
in I-III can vary from a nearly folded (fully eclipsed, cis) orientation XI to a more extended partially eclipsed (anticlinal)
conformation XII. The observation that rigid molecules, e.g.,
several tetrahydroisoquinolines, such as (-)-1,2-dihydroxyapor-
phine (56) and (S)-salsolinol (57), that incorporate a DA framework in a rigidly fixed cis orientation of the catechol ring and
basic N lack DA receptor agonist activity points to the more extended conformation XII of I-III as the one most likely to be
involved in an interaction with the receptor. In this conformation the distance between either of the catecholic OHs, both of

which seem necessary for D-1 receptor agonist activity, and the
N is about 7.0 Å.

a, X = Y = R = H   b, X = Cl, Y = OH, R = H

c, X = Cl, Y = OH, R = CH₂CH=CH₂

Discussion

Most DA receptor agonists can be categorized (58) as (1)
phenethylamines (e.g., 59, 60, 61), (2) aporphines (e.g., 62-65),
(3) aminotetralins (e.g., 66-70), (4) aminoindans (e.g., 71-74),
(5) octahydrobenzo(f)quinolines (44), (6) octahydrobenzo(g)quin-
olines (46, 75), (7) phenanthrenes (76), (8) imidazolines (77),
(9) piperazines (78), and (10) ergolines (79-83). The tetrahy-
dro-3-benzazepines (5-10), such as I-III, represent a relatively
recent addition to the list.

The structural features required for binding at brain DA
receptor (D₂, [³H]-neuroleptic labelled) sites have been summarized (58). These are: (1) a hydrogen bonding group at a
position equivalent to the meta-OH of DA and possibly an accessory hydrogen bonding site at a position complementary to
the para-OH, (2) non-essential, but potency enhancing, fat solu-
bility, (3) a basic N located about 0.6 Å from the plane of the
ring bearing the hydrogen bonding groups, (4) a distance of less
than 7.3 Å between the OH (or OH-simulating) group and the basic
N (a minimum distance is not established, but it is interesting
to note that 4-hydroxy-2-dipropylaminoindan (73) which has an N
to O distance of 5.5 Å is more potent than its 5-OH isomer that
has a greater N to O distance), and (5) steric hindrance factors
or sites of bulk intolerance, one of which is located near the N
binding site that would hinder accommodation of a tilted ring as
in octahydrobenzo(g)quinolines and groups larger than propyl in
N-substituted norapomorphine (46) and another, equivalent to the
site of steric hindrance in the model depicted in Figure 2, that
interferes with attachment of octahydrobenzo(h)quinolines (84),
isoapomorphine (85) and appropriately substituted dihydrophenan-
threnes (76). Even these detailed structural requirements are
met in some compounds, namely, properly substituted piperidines
(61, 86), e.g., 2-(3,4-dihydroxybenzyl)piperidine, and related

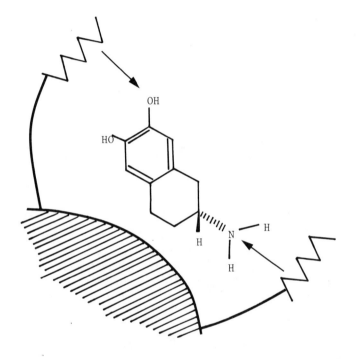

Figure 2.   *Hypothetical binding of (R)-2-amino-6,7-dihydroxy-1,2,3,4-tetrahydro-*
naphthalene to postulated DA receptor (17).

1-(3,4-dihydroxybenzyl)tetrahydroisoquinolines (61), that lack
significant DA-like activity. One of these, N-(n-propyl)-3-(3-
hydroxyphenyl)piperidine (3-PPP) is a potent agonist of DA autoreceptors (86) suggesting that some DA receptor subtypes may have
different structural requirements.

The general SAR requirements summarized for DA receptor
agonists (58), as well as consideration of the steric requirements of dopaminergic drugs (87), have led to suggestions of
various models to rationalize the interaction of these agents
with the receptor. Many of these models (e.g., 88-91) suggest
modes of receptor binding and steric parameters. They generally
suggest a steric site on the receptor that resists bulk on the
part of the DA agonist molecule on the side opposite the OH that
is "meta" to the ethylamine side chain. These studies, however,
are of little application in rationalizing enantioselectivity.

Apomorphine is enantioselective; the "natural" (-)-enanti-
omer, having the absolute 6aR stereochemistry, has dopaminergic
activity (15). This enantioselectivity of DA receptors was later
confirmed by resolution, determination of absolute stereochemistry and study of a number of hydroxyl-substituted 2-amino-1,2-
3,4-tetrahydronapthalene derivatives (18, 48, 94). Information
obtained from these studies was employed by McDermed (17) for
derivation of an appealingly simple schematic model consistent
with observed enantioselectivity. This model suggests receptor
interaction with the OH group "meta" to the ethylamine side chain
in the DA framework, a properly oriented basic amine and a region
of bulk intolerance as illustrated in Figure 2 for (R)-2-amino-6,
7-dihydroxy-1,2,3,4-tetrahydronaphthalene (ADTN). The site of
steric hindrance, indicated by the lined area in Figure 2 (17)
was supported by examination of 5- and 8-propyl derivatives; as
expected the 8-, but not the 5-isomer, was a fairly potent DA
receptor agonist (19).

The benzazepines I-III are not readily incorporated into
this model for several reasons. Firstly, both of the catecholic
OHs in I-III are "meta" to an ethylamine chain and both are essential for activity (5, 35). In addition, either the 6-Cl or
the 1-phenyl substituent must reside near the postulated site of
bulk intolerance (51). As illustrated in Figures 3 and 4 (36),
at least two plausible interpretations of the enantioselectivity
of I-III may be advanced to accommodate this model. In these
Figures, the possibility is also advanced that receptor interaction might conceivably involve the p orbitals of an OH group.
As indicated in Figure 3, the phenyl or substituted phenyl group
of (R)-I-III might be accommodated as a result of its being in
an α-orientation that avoids the site of steric hindrance. An
alternative possibility, suggested in Figure 4, is that interaction might involve the N, the N-allyl (except in the case of (S)-
III, in which case this group may meet a secondary site of bulk
intolerance, suggested by Seeman (58) as being near the N-binding
site), the 6-chloro, the catecholic OHs, and the phenyl or 4'-

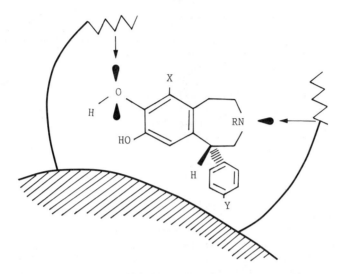

Figure 3. Possible mode of binding of (R)-I-III to postulated DA receptor. (Reproduced with permission from Ref. 36. Copyright, Swedish Academy of Pharmaceutical Science.)

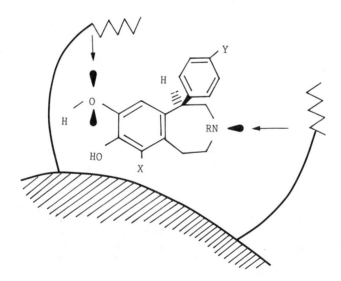

Figure 4. Alternative possible mode of binding of (R)-I-III to postulated DA receptor. (Reproduced with permission from Ref. 36. Copyright, Swedish Academy of Pharmaceutical Science.)

hydroxyphenyl, all of which, as noted previously, increase some kinds of DA-like activity. A preference for this latter interpretation is suggested by several other DA receptor models, alluded to in the subsequent discussion, that postulate an accessory binding site for the phenyl group in this general location. Possibly as a result of the requirement of both OHs, the mode of interaction of the benzazepines with DA receptors is somewhat altered so that Cl substituent can fit the receptor and even provide enhanced binding, perhaps by increasing the acidity of the adjacent OH (39), and placing the 1-phenyl group in the correct configurational position to favor binding with postulated auxiliary lipophilic binding site. This concept is favored by the negative effect on in vitro DA-like potency noted upon introduction of a hydrophilic 4-OH on the 1-phenyl substituent; i.e., a modification that would be detrimental to supplemental binding to the accessory lipophilic site.

Another topographical model, advanced for the renal vascular DA receptor by Erhardt (92) is described in greater detail in another section of this monograph (93). This model locates important receptor sites on Cartesian coordinates. It extends the McDermed model by suggesting a second site of steric hindrance about 2.0 Å above the plane of the ethylamine chain and an auxiliary binding site, alluded to previously, opposite the principal site of bulk intolerance. As this model, which is consistent with the structures of most DA receptor agonists, specifically locates the amine and "meta"-OH it can be utilized to rationalize the enantioselectivity of known chiral DA receptor agonists (94).

Considering that the benzazepines I-III bind with the receptor so that the 1-phenyl, or 1-hydroxyphenyl, substitutent locates near the auxiliary site and viewing the benzazepines in a reasonable energy conformation (95) through the azepine ring toward the edge of the catechol ring shows that in the R isomers XIII the phenyl, in a preferred equatorial orientation, assumes a location slightly above the auxiliary binding site, whereas the S enantiomers XIV place the 1-substituent slightly below (or into) this site. This may explain the enantioselectivity of these compounds.

Still another DA receptor model, based on the study of aporphines, takes chirality factors into consideration. This model suggested by Neumeyer (16, 96) postulates an obstruction or "obstacle" on the receptor that precludes appropriate interaction of aporphines in the 6aS configuration with the receptor. This model, in common with many others, also takes into account OH binding sites, conformational aspects, steric hindrance factors, and N-substitution with the observation as noted elsewhere (e.g., 97), that N-substitution with groups as large as propyl may provide supplemental binding to favor D-2 receptor interaction. As the asymmetric center of apomophine is located on the carbon α to the N, whereas in I-III it is located on the ben-

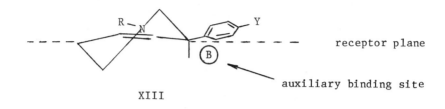

XIII

receptor plane

auxiliary binding site

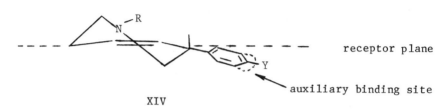

XIV

receptor plane

auxiliary binding site

zylic carbon this model is not predictive of the enantioselectivity of the dopaminergic benzazepines. It is noteworthy, however, that the benzazepines are accommodated by the model and
when the catecholic OHs and the basic N groups are approximately
superimposed with the 1-substituent in a preferred equatorial
orientation this substituent is located in about the same vicinity as the A ring of apomorphine. This is the same location as
that suggested for the auxiliary binding site in the Erhardt
model (92, 93).

Many antipsychotics apparently act by direct or functional
blockade of DA receptors in the brain. Since the discovery of
the first antipsychotic drugs in the early 1950s, a great number
of structurally diverse compounds that produce similar pharmacological or biochemical actions, the neuroleptics, has been discovered (98). For many of the neuroleptics, there exists convincing evidence that they antagonize DA receptors. Unfortunately, diversity of the chemical structures of neuroleptic DA
receptor antagonists, lack of uniformity in the assay procedures
identifying them, possibly differing modes of action (99), interaction with multiple or different receptors (100, 101), metabolic transformations in vivo, etc., hinder application of the
enormous amount of data accumulated for these agents to identification of their pharmacophoric constituents (98). Additionally, many DA receptor antagonists are so conformationally flexible that even the many x-ray, NMR and quantum chemical studies
of these agents are of limited value in identification of their
conformation during receptor interaction (102). Thus, the veritable wealth of SAR data for neuroleptics remains a virtually

untapped source of information in defining the binding sites and
topography of DA receptors, although a few studies have related
some antagonist structures to that of the neurotransmitter (49,
103, 104). In utilizing SAR information derived from DA receptor antagonists for identifying agonist pharmacophores considerable caution is suggested (19). A great deal of evidence indicates that agonist and antagonist sites may have different properties (20, 58, 105, 106, 107).

For several neuroleptics, e.g., butaclamol and related compounds (20-23), sulpiride (24), octoclothepin (25-28), many
aminoalkylated tricyclics (29), etc., optical antipodes have
been studied. Enantioselectivity is invariably observed (108-
113). Butaclamol (XV) and some of its analogs, which are rigid
structures with a high degree of enantiospecificity, have been
utilized by Humber and his associates (20, 21, 22) for derivation of a three-dimensional map of the receptor for these DA
antagonists. As this map was derived by the observation of
conformational and configurational similarities in the phene-
thylamine part of the rigid DA receptor agonist apomorphine and
the analogous portion of butaclamol, it is reasonable, but not
necessarily required, to expect that it also applies to DA agonist recognition sites. Important features of the three-dimensional map of the receptor for the DA antagonistic 3S, 4aS, 13bS-
butaclamol (XV) are location on Cartesian coordinates of dimen-
sionally-defined binding sites. These are primary binding sites
occupied by the phenyl ring A and the basic N (located 0.9 Å
above the plane ring A) (20, 22) and an accessory binding site
that accommodates the t-butyl group of XV at a location vectori-
ally defined relative to the N (21).

XV   XVI

It is noteworthy that the DA receptor agonist benzazepines
I-III, as illustrated by the general structure XVI (particularly
strikingly in a conformationally disfavored pseudo chair form),
are capable of occupying the primary binding sites defined in the
butaclamol-derived map. Thus, the o-(OH)$_2$-substituted phenyl ring
and N of XVI are almost perfectly superimposable upon ring A and
the basic N of XV. Clearly the N of XIV can assume a position of
0.9 Å above the plane of the catechol ring. Perhaps most signi-

ficantly the absolute stereochemistry at position 1 of (R)-XVI
is identical with that of (13bS)-XV. Additionally the 1-phenyl
group in this orientation is able to attain a rotameric conformation that permits almost perfect superimposition with the C
ring of conformer B (22) of XV. It should be noted that the
overall conformation of the tetrahydrobenzazepine nucleus and
the rotameric form of the 1-phenyl substituent of XVI suggested
for optimal fit into this map differs from that postulated for
interaction with DA receptor models derived from agonists. Conceivably, this less preferred conformation may account for the
DA-antagonist activity noted for I. This speculation is consistent with the observation (7) that removal of the 1-phenyl substituent (with appropriate relocation), a change that would favor attainment of the conformationally less favored pseudo chair
form of the azepine ring, results in accentuation of DA receptor
antagonist activity.

The another hypothetical three-dimensional model based on the action of neuroleptics, e.g., butaclamol, and the DA receptor agonist apomorphine is described in another section of this monograph by Olson and his associates (97, 114). It is based on
plausible interactions between pharmacophoric groups of structurally diverse neuroleptics and postulated amino acid side
chain substituents of the receptor protein. Three essential
binding sites (one possibly needed for antagonism) and one
auxiliary lipophilic site are suggested. Derivation of the
model assumes that within a given class, such as different DA
receptors, the same segments of the protein bind drugs whose
action is mediated by that receptor. This is consistent with
the general observation (19) that DA receptors in different anatomical locations and of apparently distinct pharmacological
types seem to have comparable stereochemical features. These
binding regions are complementary with ring A of butaclamol
(XV), to ring C (perhaps required for antagonism), the protonated amino group, and a fourth binding region of low structural
specificity corresponding to the auxiliary binding site identified in the butaclamol map. In the model, geometry is defined
via three-dimensional structures of drugs that modulate DA receptors. This model, can clearly be accommodated, as described
for the conformation XVI in the Humber map (20, 21, 22), most
advantageously by the R enantiomers of I-III. Interestingly the
site associated with DA receptor antagonism is accommodated by
the 1-substituent of the tetrahydrobenzazepines. This may offer
a rationale for the partial DA antagonist activity of some of
these compounds (2).

The observation that the polypeptide neurotensin has pharmacological properties similar to those of neuroleptics (115)
has led to another hypothetical structure of the DA receptor.
This theoretical receptor that considers neurotensin as part of
the DA receptor accommodates DA agonists and antagonists of diverse structure (116). It is suited to the dopaminergic 3-ben-
zazepines, but it is not predictive of enantioselectivity.

Summary and Conclusions

The results of the present study indicate that the 1-phenyl substituted tetrahydro-3-benzazepines I-III interact with DA receptors in a stereoselective manner. Enantioselectivity favors the R configuration, especially in in vitro tests. As the 1-phenyl group contributes significantly to receptor interaction it likely binds to an accessory site. Comparison of the probable binding modes of the dopaminergic benzazepines with various hypothetical models uniformly suggests the molecules can be accommodated by receptor interaction with primary binding sites complementary to the catecholic OH groups and a N functionality. The various models generally suggest that the benzazepines bind in a fashion slightly different from that of DA and related structures that can achieve a nearly extended orientation between the catechol ring and the N. Allylation of the N in the appropriate vectorial direction enhances D-2 receptor agonist activity. Perhaps the vectorial direction of the lone pair of electrons on the N is important for DA agonist activity. This is particularly important in agents that incorporate the N in a fixed (ring) position. Most models suggest the accessory binding site that binds the phenyl ring is on the side of the receptor opposite a commonly postulated site of bulk intolerance. In DA receptor models derived from antagonists the 3-benzazepines must assume a different conformation for superimposition of the suggested sites of binding of the catechol ring, the N and the 1-phenyl substituent. Possibly the 3-benzazepines owe their agonist activity to interaction with the receptor in one conformation whereas their antagonist activity derives from receptor accommodation of a second, less preferred conformation. In agreement with previous observations (117), it appears that no single model rationalizes agonist activity of all classes of dopaminergic agents.

Literature Cited

1. Bergin, R.; Carlstrom, D. Acta Crystallogr. 1968, B24, 1506.
2. Setler, P.E.; Sarau, H.M.; Zirkle, C.L.; Saunders, H.L. Eur. J. Pharmacol. 1978, 50, 419.
3. Pendleton, R.G.; Samler, L.; Kaiser, C.; Ridley, P.T. Eur. J. Pharmacol. 1978, 51, 19.
4. Kebabian, J.W.; Calne, D.B. Nature (London) 1979, 277, 93.
5. Wilson, J.W. "Program and Abstracts", National Medicinal Chemistry Symposium of the American Chemical Society, 16th, Kalamazoo, MI, June 18-22, 1978; American Chemical Society: Washington, D.C., 1978, p. 155.

6. Weinstock, J.; Wilson, J.W.; Ladd, D.L.; Brush, C.K.;
    Pfeiffer, F.R.; Kuo, G.Y.; Holden, K.G.; Yim, N.C.F.;
    Hahn, R.A.; Wardell, J.R., Jr.; Tobia, A.J.; Setler, P.E.;
    Sarau, H.M.; Ridley, P.T. J. Med. Chem. 1980, 23, 973.
7. Kaiser, C.; Ali, F.E.; Bondinell, W.E.; Brenner, M.;
    Holden, K.G.; Ku, T.W.; Oh, H.-J.; Ross, S.T.; Yim,
    N.C.F.; Zirkle, C.L.; Hahn, R.A.; Sarau, H.M.; Setler,
    P.E.; Wardell, J.R., Jr. J. Med. Chem. 1980, 23, 975.
8. Hahn, R.A.; Wardell, J.R., Jr.; Sarau, H.M.; Ridley, P.T.
    J. Pharmacol. Exp. Ther. 1982; in press.
9. Stote, R.M.; Erb, B.; Alexander, F.; Givens, K.; Familiar,
    R.; Dubb, J. Kidney Int. 1982, 21, 248.
10. Blumberg, A.L.; Hieble, J.P.; McCafferty, J.; Hahn, R.A.;
    Smith, J., Jr. Fed. Proc., Fed. Am. Soc. Exp. Biol. 1982,
    41, 1345.
11. Sibley, D.R.; Leff, S.E.; Creese, I. Life Sci. 1982; in
    press.
12. Ackerman, D.M.; Weinstock, J.; Wiebelhaus, V.D.; Berkowitz,
    B.A. Drug Dev. Res. 1982, 2, 283.
13. Sarau, H.M.; unpublished results.
14. Setler, P.E.; unpublished results.
15. (a) Saari, W.; King, S.W.; Lotti, V.J. J. Med. Chem.
    1973, 16, 171; (b) Neumeyer, J.L.; Neustadt, B.R.; Oh,
    K.H.; Weinhardt, K.K.; Boyce, C.B.; Rosenberg, F.J.;
    Teiger, D.G. J. Med. Chem. 1973, 16, 1223.
16. Neumeyer, J.L.; Law, S.J.; Lamont, J.S. "Apomorphine and
    Other Dopaminomimetics, Vol. 1: Basic Pharmacology";
    Gessa, G.L.; Corsini, G.U., Eds.; Raven Press; New York,
    1981; 209.
17. McDermed, J.D.; Freeman, H.S.; Ferris, R.M. "Catechol-
    amines: Basic and Clinical Frontiers"; Usdin, E.; Kopin,
    I.J.; Barchas, J., Eds.; Pergamon; New York, 1978, pp.
    568-570.
18. Tedesco, J.L.; Seeman, P.; McDermed, J.D. Mol. Pharmacol.
    1979, 16, 369.
19. McDermed, J.D.; Freeman, H.S. "Symposium on Dopamine
    Receptor Agonists", Stockholm, Sweden, Abstracts, April
    20-23, 1982.
20. Humber, L.G.; Bruderlein, F.T.; Voith, K. Mol. Pharmacol.
    1975, 11, 833.
21. Humber, L.G.; Bruderlein, F.T.; Philipp, A.H.; Götz, M.;
    Voith, K. J. Med. Chem. 1979, 22, 761.
22. Philipp, A.H.; Humber, L.G.; Voith, K. J. Med. Chem.
    1979, 22, 768.
23. Bird, P.; Bruderlein, F.T.; Humber, L.G. Can. J. Chem.
    1976, 54, 2715.
24. Jenner, P.; Clow, A.; Reavill, C.; Theodorou, A.; Marsden,
    C.D. J. Pharm. Pharmacol. 1980, 32, 39.
25. Metysova J.; Protiva, M. Act. Nerv. Super. 1975, 17, 218.
26. Petcher, T.J.; Schmutz, J.; Weber, H.P.; White, T.G.
    Experientia 1975, 31, 1389.

27.   Seidlova, V.; Protiva, M.   Collect. Czech. Chem. Commun.
      1967, 32, 1747.
28.   Jaunin, A.; Petcher, T.J.; Weber, H.P.   J. Chem. Soc.,
      Perkin II   1977, 186.
29.   Kaiser, C.; Zirkle, C.L.; unpublished results.
30.   Ingoglia, N.A.; Dole, V.P.   J. Pharmacol. Exp. Ther.
      1970, 175, 84.
31.   Berkowitz, B.A.; Way, E.L.   J. Pharmacol. Exp. Ther.
      1971, 177, 500.
32.   Abdel-Monem, M.M.; Larson, D.L.; Kupferberg, H.J.;
      Portoghese, P.S.   J. Med. Chem.   1972, 15, 494.
33.   Sullivan, H.R.; Due, S.L.; McMahon, R.E.   J. Pharm.
      Pharmacol.   1975, 27, 728.
34.   Portoghese, P.S.   Acc. Chem. Res.   1978, 11, 21.
35.   Kaiser, C.; Dandridge, P.A.; Garvey, E.; Hahn, R.A.; Sarau,
      H.M.; Setler, P.E.; Bass, L.S.; Clardy, J.   J. Med. Chem.
      1982, 25, 697.
36.   Kaiser, C.; Dandridge, P.A.; Weinstock, J.; Ackerman, D.M.;
      Sarau, H.M.; Setler, P.E.; Webb, R.L.; Horodniak, J.W.;
      Matz, E.D.   Acta Pharm. Seuc.   1982; in press.
37.   Clardy, J.; unpublished results.
38.   These drawings were generated by D.E. Zacharias from
      single-crystal x-ray crystallographic data generated as
      described in reference 35, J. Clardy; unpublished results.
39.   Weinstock, J.; Wilson, J.W.; Ladd, D.L.; Brenner, M.;
      Ackerman, D.M.; Blumberg, A.L.; Hahn, R.A.; Hieble, J.P.;
      this Am. Chem. Soc. Symposium Series book.
40.   Fujita, N.; Saito, K.   Neuropharmacology   1978, 17, 1089.
41.   Frey, E.A.; Cote, T.E.; Grewe, C.W.; Kebabian, J.W.
      Endocrinology   1982, 110, 1897.
42.   Ackerman, D.M.; Blumberg, A.L.; McCafferty, J.P.; Sherman,
      S.S.; Weinstock, J.; Kaiser, C.; Berkowitz, B.   Fed. Proc.,
      Fed. Am. Soc. Exp. Biol.   1982; in press.
43.   Cannon, J.G.   Adv. Neurol.   1975, 9, 177.
44.   Cannon, J.G.   Adv. Biosci.   1978, 20, 87.
45.   Cannon, J.G.; Suarez-Gutierrez, C.; Lee, T.; Long, J.P.;
      Costall, B.; Fortune, D.H.; Naylor, R.J.   J. Med. Chem.
      1979, 22, 341.
46.   Cannon, J.G.; Lee, T.; Goldman, H.D.; Long, J.P.; Flynn,
      J.R.; Verimer, T.; Costall, B.; Naylor, R.J.   J. Med.
      Chem.   1980, 23, 1.
47.   Cannon, J.G.; Lee, T.; Goldman, H.D.; Costall, B.; Naylor,
      R.J.   J. Med. Chem.   1977, 20, 1111.
48.   McDermed, J.; McKenzie, G.M.; Freeman, H.S.   J. Med. Chem.
      1976, 19, 547.
49.   Horn, A.; Snyder, S.H.   Proc. Natl. Acad. Sci., U.S.A.
      1971, 68, 2325.
50.   Woodruff, G.N.; Watling, K.J.; Andrews, C.D.; Poat, J.A.;
      McDermed, J.D.   J. Pharm. Pharmacol.   1977, 29, 422.

51. Kotake, A.; Goldberg, L.; Hoffmann, P.; Cannon, J.
     Pharmacologist 1977, 19, 208.
52. Cannon, J.G.; Costall, B.; Laduron, P.M.; Leysen, J.E.;
     Naylor, R.J. Biochem. Pharmacol. 1978, 27, 1417.
53. Costall, B.; Lim, S.K.; Naylor, R.J.; Cannon, J.G. J.
     Pharm. Pharmacol. 1982, 34, 246.
54. Bergin, R.; Carlstrom, D. Acta Crystallogr. 1968, B24,
     1506.
55. Giessner-Prettre, C.; Pullman, B. J. Mag. Resonance 1975,
     18, 564.
56. Miller, R.J.; Kelly, P.H.; Neumeyer, J.L. Eur. J.
     Pharmacol. 1976, 35, 77.
57. Seeman, P.; Titeler, M.; Tedesco, J.; Weinrich, P.;
     Sinclair, D. Adv. Biochem. Psychopharmacol. 1978, 19,
     167.
58. Seeman, P. Pharmacol. Rev. 1980, 32, 229.
59. Borgman, R.J.; McPhillips, J.J.; Stitzel, R.E.; Goodman,
     I.J. J. Med. Chem. 1973, 16, 630.
60. Costall, B.; Naylor, R.J.; Pinder, R.M. J. Pharm.
     Pharmacol. 1974, 26, 753.
61. Ginos, J.Z.; Cotzias, G.C.; Tolosa, E.; Tang, L.C.;
     LoMonte, A. J. Med. Chem. 1975, 18, 1194.
62. Lal, S.; Sourkes, T.L.; Missala, K.; Belendiuk, G. Eur.
     J. Pharmacol. 1972, 20, 71.
63. Neumeyer, J.L.; Granchelli, F.E.; Fuxe, K.; Ungerstedt,
     U.; Corrodi, H. J. Med. Chem. 1974, 17, 1090.
64. Pearl, J. J. Pharm. Pharmacol. 1978, 30, 118.
65. Sheppard, H.; Burghardt, C.R. Biochem. Pharmacol. 1978,
     27, 1113.
66. Ilhan, M.; Kitzen, J.M.; Cannon, J.G.; Long, J.P. Eur. J.
     Pharmacol. 1977, 41, 301.
67. Burkman, A.M. Neuropharmacology 1973, 12, 83.
68. Horn, A.S. J. Pharm. Pharmacol. 1974, 26, 735.
69. McDermed, J.D.; McKenzie, G.M.; Phillips, A.M. J. Med.
     Chem. 1975, 18, 362.
70. Woodruff, G.N.; Watling, K.J.; Andrews, C.D.; Poat, J.A.;
     McDermed, J.D. J. Pharm. Pharmacol. 1977, 29, 422.
71. Cheng, H.-C.; Long, J.P.; van Orden, L.S., III; Cannon,
     J.G.; O'Donnell, J.P. Res. Commun. Chem. Pathol.
     Pharmacol. 1976, 15, 89.
72. Rusterholz, D.B.; Long, J.P.; Flynn, J.R.; Cannon, J.G.;
     Lee, T.; Pease, J.P.; Clemens, J.A.; Wong, D.T.; Bymaster,
     F.P. Eur. J. Pharmacol. 1979, 55, 73.
73. Hacksell, U.; Arvidsson, L.-E.; Svensson, U.; Nilsson,
     J.L.G.; Wikström, H.; Lindberg, P.; Sanchez, D.; Hjorth,
     S.; Carlsson, A.; Paalzow, L. J. Med. Chem. 1981, 24,
     429.
74. Sindelar, R.D.; Mott, J.; Barfneckt, C.F.; Arneric, S.P.;
     Flynn, J.R.; Long, J.P.; Bhatnagar, R.K. J. Med. Chem.
     1982, 25, 858.

75. Smissman, E.E.; El-Antably, S.; Hedrick, L.W.; Walaszek, E.J.; Tseng, L.-F. J. Med. Chem. 1973, 16, 109.
76. Nichols, D.E.; Toth, J.E.; Kohli, J.D.; Kotake, C.K. J. Med. Chem. 1978, 21, 395.
77. Struyker-Boudier, H.; Teppema, L.; Cools, A.; Van Rossum, J. J. Pharm. Pharmacol. 1975, 27, 882.
78. Volkman, P.H.; Goldberg, L.I. Pharmacologist 1976, 18, 130.
79. Camerman, N.; Chan, L.Y.Y.; Camerman, A. Mol. Pharmacol. 1979, 16, 729.
80. Camerman, N.; Camerman, A. Mol. Pharmacol. 1981, 19, 517.
81. Bach, N.J.; Kornfeld, E.C.; Jones, N.D.; Chaney, M.O.; Dorman, D.E.; Paschal, J.W.; Clemens, J.A.; Smalstig, E.B. J. Med. Chem. 1980, 23, 481.
82. Cannon, J.G.; Demopoulos, B.J.; Long, J.P.; Flynn, J.R.; Sharabi, F.M. J. Med. Chem. 1981, 24, 238.
83. Euvard, C.; Ferland, L.; Fortin, M.; Oberlander, C.; Labrie, F.; Boissier, J.R. Drug Dev. Res. 1981, 1, 151.
84. Cannon, J.G.; Lee, T., Hsu, F.-L, Long, J.P.; Flynn, J.R. J. Med. Chem. 1980, 23, 502.
85. Neumeyer, J.L.; McCarthy, M.; Battista, S.P.; Rosenberg, F.J.; Teiger, D.G. J. Med. Chem. 1973, 16, 1228.
86. Hjorth, S.; Carlsson, A.; Lindberg, P.; Sanchez, D., Wikström, H.; Arvidsson, L.-E.; Hacksell, U.; Nilsson, J.L.G.; Svensson, U. Proc. Am. College of Neuropsychopharmacology, San Juan, 1979.
87. Miller, D.D. Fed. Proc., Fed. Am. Soc. Exp. Biol. 1978, 37, 2392.
88. Clement-Cormier, Y.C.; Meyerson, L.R.; Phillips, H.; Davis, V.E. Biochem. Pharmacol. 1979, 28, 3123.
89. Goldberg, L.I.; Kohli, J.D.; Kotake, A.N.; Volkman, P.H. Fed. Proc., Fed. Am. Soc. Exp. Biol. 1978, 37, 2396.
90. Sheppard, H.; Burghardt, C.R. Mol. Pharmacol. 1974, 10, 721.
91. Grol, C.J.; Rollema, H. J. Pharm. Pharmacol. 1977, 29, 153.
92. Erhardt, P.W. J. Pharm. Sci. 1980, 69, 1059.
93. Erhardt, P.W.; this Am. Chem. Soc. Symposium Series book.
94. Erhardt, P.W. Acta Pharm. Seuc. 1982; in press.
95. Sauriol-Lord, F.; Grindley, T.B. J. Am. Chem. Soc. 1981, 103, 936.
96. Neumeyer, J.L.; Arana, G.W.; Ram, V.J.; Baldessarini, R.J. Acta Pharm. Seuc. 1982; in press.
97. Olson, G.L.; Cheung, H.-C.; Chiang, E.; Berger, L.; this Am. Chem. Soc. Symposium Series book.
98. Kaiser, C.; Setler, P.E. "Burger's Medicinal Chemistry"; 4th ed., Wolff, M.E., Ed.; Wiley, New York, 1982, Vol. 3, Chapter 56. pp. 859-980.
99. Van Rossum, J.M. Fed. Proc., Fed. Am. Soc. Exp. Biol. 1978, 37, 2415.

100. Leysen, J.E.; Gommeren, W.; Laduron, P.M. Biochem. Pharmacol. 1978, 27, 307.
101. Leysen, J.E.; Niemegeers, C.J.E.; Tollenaere, J.P.; Laduron, P.M. Nature (London) 1978, 272, 168.
102. Tollenaere, J.P.; Moereels, H.; Koch, M.H.J. Eur. J. Med. Chem. 1977, 12, 199.
103. Feinberg, A.P.; Snyder, S.H. Proc. Natl. Acad. Sci., U.S.A. 1975, 72, 1899.
104. Janssen, P.A.J. "Psychopharmacological Agents"; Gordon, M., Ed.; Academic Press, New York, 1974, Vol. 3, p. 129.
105. Creese, I.; Burt, D.R.; Snyder, S.H. Life Sci., 1975, 17, 993.
106. Lew, J.Y.; Goldstein, M. Eur. J. Pharmacol. 1979, 55, 429.
107. Suen, E.T.; Stefanini, E.; Clement-Courmier, Y.C. Biochem. Biophys. Res. Commun. 1980, 96, 953.
108. Pelz, K.; Svatek, E.; Metysova, J.; Hradil, F.; Protiva, M. Collect. Czech. Chem. Commun. 1970, 35, 2623.
109. Engelhardt, E.L.; Christy, M.E.; Colton, C.D.; Freedman, M.B.; Boland, C.C.; Halpern, L.M.; Vernier, V.G.; Stone, C.A. J. Med. Chem. 1968, 11, 325.
110. Moller-Nielsen, I.; Hougs, W.; Lassen, N.; Holm, T.; Peterson, P. Acta Pharmacol. Toxicol. 1962, 19, 87.
111. Garan, L.; Govoni, S.; Stephanini, E.; Trabuchi, M.; Spano, P.F. Life Sci. 1978, 23, 1745.
112. Seeman, P.; Westman, K.; Protiva, M.; Jilek, J.; Jain, P.C.; Saxena, A.K.; Amand, N.; Humber, L.; Philipp, A.H. Eur. J. Pharmacol. 1979, 56, 247.
113. Remy, D.C.; Rittle, K.E.; Hunt, C.A.; Anderson, P.S.; Arison, B.H.; Engelhardt, E.L.; Hirschmann, R.; Clineschmidt, B.V.; Lotti, V.J.; Bunting, P.R.; Ballantine, R.J.; Papp, N.L.; Flataker, L.; Witoslawski, J.J.; Stone, C.A. J. Med. Chem. 1977, 20, 1013.
114. Olson, G.L.; Cheung, H.-C.; Morgan, K.D.; Blount, J.F.; Todaro, L.; Berger, L.; Davidson, A.B.; Boff, E. J. Med. Chem. 1981, 24, 1026.
115. Ervin, G.N.; Birkemo, L.S.; Nemeroff, C.B.; Prange, A.J., Jr. Nature (London) 1981, 291, 73.
116. Smythies, J.R. Med. Hypotheses 1981, 7, 1449.
117. Cannon, J.G.; Long, J.P.; Bhatnagar, R. J. Med. Chem. 1981, 24, 1113.

RECEIVED November 18, 1982

Commentary: Stereoisomeric Probes of the Dopamine Receptor

JOHN McDERMED

Wellcome Research Laboratories, Department of Organic Chemistry,
Research Triangle Park, NC 27709

Until very recently apomorphine (APO) was the only important chiral DA agonist whose enantiomers had been studied. The
only congeneric groups of resolved agonists to be reported subsequently were 2-aminotetralins (1,2). The present enantiomer
study of the SK&F 38393 series thus provides welcome new data
for stereochemical definition of DA receptors. Moreover, the
previous studies employed rather narrow choices of pharmacological models. The broader range of models presented here is
therefore particularly helpful.

The discussed correlation between pharmacological profiles
and absolute configurations needs little further comment.
Qualitatively coherent enantioselectivities in the series
generally support the dopaminergic character of the models.
Only the unexpected result with (R)- and (S)-III is remarkable.
The fact that (R)-III is 30 times more potent than (R,S)-III in
vivo as a renal vasodilator is not reflected in vitro in the
other models. A possible change in mechanism of action was
mentioned. The effects of (R)- and (S)-III on α-adrenergic
receptors, for example, are unreported but could be relevant.
The data may not rule out receptor blockade by (S)-III (e.g.,
what is the effect of (S)-III on the activities of other agonists such as DA or (R)-II?), though the binding data offer no
hint of that. This intriguing observation merits further
study.

The possible relationship between the structures and absolute configurations of butaclamol and agonists of the present
series (XV and XVI) is interesting. Analogously, a possible
relationship between the configurations of (+)-butaclamol and
(-)-APO has been repeatedly remarked and can be construed as
evidence for the hypothesis that the stereochemistry of DA
agonist recognition sites should be inferable by analysis of
the structures of such familiar DA antagonists (3). However,
the appropriateness of such comparisons is equivocal. It is a
widely overlooked fact that bulbocapnine is a DA antagonist of
the aporphine group with absolute stereochemistry opposite to

0097-6156/83/0224-0247$06.00/0

that of (-)-APO (4). Moreover, it has lately been reported
that (+)-APO is an effective, if relatively weak, DA antagonist
(5). The implications of these examples of inversion of
stereochemical demand are far from clear, but should signal the
risk of intuitive comparisons of agonists and antagonists.
Elucidation of the enantioselectivities of the DA antagonists
of the SK&F 38393 series would be particularly interesting in
this connection.

Kaiser refers to several receptor models that have been
devised to account for SAR in other groups of dopaminergic
ligands. Those specifically based on agonists all include most
fundamentally a primary binding site for a basic N atom and a
primary binding site for at least one hydroxyl group, in a more
or less precisely defined spatial relationship to each other.
Secondarily, they diversely postulate auxiliary binding sites
and/or points of steric occlusion in regions relatable to the
primary binding sites. It is fair to say, however, that moderately precise specification of even those most fundamental
sites introduces confounding conundrums. For example, why is
the very short O-N distance in 4-hydroxy-2-aminoindans apparently acceptable (6), and why is a single "meta" hydroxyl
sufficient for activity in the aporphines and their partial
structures but not in the agonists of the SK&F 38393 series?
It should be added that no experimental evidence exists
providing a molecular explanation for the requirements and role
of the aromatic hydroxyl(s). Descriptively, it is clear that a
hydroxyl "meta" to an aminoethyl fragment is important, but
almost nothing is known about the cause of this "meta effect".
Only the recent demonstration that this effect is chirally
modulated (1) suggests that factors extrinsic to the drug
(i.e., specific drug-receptor interactions), rather than purely
intrinsic factors (e.g., O-N distance or drug dipoles), may be
responsible.

In the present series the catechol hydroxyls are displayed
symmetrically with respect to the N atom, and both appear to be
required for agonist activity. Thus the present study does not
appear to clarify the role of the hydroxyls in agonist-receptor
interactions. However, Kaiser's unelaborated conjecture that
binding to one of the hydroxyls might be effected through its
unshared electrons is provocative and potentially significant,
though very difficult to address experimentally. From X-ray
studies of enzyme-inhibitor complexes there is precedent for
imagining that the hydroxyl binding site may not be
identifiable as a discrete functional group in a fixed
location. Rather, it could be a more complex environment
involving molecules of fixed water (7) in which array the
agonist hydroxyl plays a key organizational role, via multiple
hydrogen bonding, during the process of receptor binding or
activation or both. In such a circumstance the effective
ligand would actually be the drug plus its array of fixed

water, so that consideration of the structure of the drug alone
could be misleading. Definitive evidence about such a scheme
would only be available from high resolution X-ray crys-
tallography.

The one steric exclusion zone included in Figures 3 and 4
was not very helpful in rationalizing the observed
enantioselectivity. Thus, the present study is silent on the
validity of this and all of the several other loci of steric
bulk postulated in different receptor models. On this point it
would be prudent to reflect that without exception these steric
features have been invented to rationalize the inactivities of
a very few compounds (sometimes single ones) which were perceived to be analogs of active agonists modified by addition of
steric bulk. This syllogism minimally requires the treacherous
assumption that the relevant conformational and electronic
properties of the active agonist are unchanged by this addition
of steric bulk or conformational restraint. Embrace of all
such steric features of receptors should be very cautious.

One receptor feature which is significantly addressed by
the present study is the possible existence of an accessory
binding site for the 1-phenyl substituent, as in XIII. The
location of this auxiliary binding site seems to be in the same
region as that which would accommodate the unsubstituted ring
of APO. Interestingly, available data from aporphine partial
structures argue **against** a positive contribution to binding
from this ring. In the SK&F 38393 series the major contribution to potency and the enantioselectivity attributable to the
1-phenyl group are entirely consistent with the auxiliary binding site hypothesis. However, it is alternatively plausible
that potency enhancement and enantioselectivity could arise
from purely intrinsic conformational effects and need not
necessarily imply auxiliary binding. The 7-membered ring
enjoys great conformational mobility permitting the
all-important N atom and its electrons a wide range of possible
locations and orientations. A major role of the 1-phenyl group
could be the stabilization of one particular conformation of
the heterocyclic ring which is appropriate for receptor
binding, thus enhancing potency. Any such asymmetric perturba-
tion at the 1-position would produce a chiral distortion of the
ring and could give rise to enantioselectivity. Consideration
of substituent effects on the 1-phenyl group might help differentiate this possibility.

Some of the structural properties which have been ascribed
to DA receptors appear to deserve attention for their heuristic
value, but painfully few should engender much confidence in
their reality. A sobering lesson is available from analysis of
complexes of dihydrofolate reductase (7,8). Methotrexate is a
very close analog of folic acid and is a potent inhibitor of
the enzyme, but it is now almost certain that these ligands
bind in the enzyme active site in aspects differing by a rota-

tion of 180°! Inscrutable opportunities for error attend
extrapolation from drug structure to binding mode and on to
receptor features. Still, enantiomer study is a resource of
exceptional subtlety for definition of pharmacological mechanisms, identification of pharmacophores, and for constructing
working hypotheses and models for drug design. The lingering
controversy over the pharmacophore of dopaminergic ergots would
no doubt be partly illuminated by resolution of some key ergot
partial structures (9) and of some recently reported ergot-
aporphine hybrids (10). Similar studies of the antagonists of
the SK&F 38393 series could be expected to bring us closer to
understanding the molecular and mechanistic relationships
between DA agonists and antagonists.

Literature Cited

1. Freeman, H.S.; McDermed, J.; in "Chemical Regulation of
    Biological Mechanisms"; Spec. Publ. 42; Creighton, A.M.;
    Turner, S., Eds.; Royal Society of Chemistry; London,
    1982; pp. 154-166.
2. Seiler, M.P.; Markstein, R.; Mol. Pharmacol. 1982, 22,
    281.
3. Humber, L.G.; Bruderlein, F.T.; Voith, K.; Mol. Pharmacol.
    1975, 11, 833.
4. Wardell, J.R., Jr.; Hahn, R.A.; Stefankiewicz, J.S.; in
    "Peripheral Dopaminergic Receptors"; Imbs, J.; Schwartz,
    J.; Eds.; Pergamon; New York, 1979; pp. 389-399.
5. Riffee, W.H.; Wilcox, R.E.; Smith, R.V.; Davis, P.J.;
    Brubaker, A.; in "Proceedings of International Dopamine
    Symposium, Okayama, Japan"; Kohsaka, M.; Shohmori, T.;
    Tsukada, Y.; Woodruff, G.N.; Eds.; Pergamon; London, 1982;
    pp. 357-362.
6. Hacksell, U.; Arvidsson, L.; Svensson, U.; Nilsson,
    J.L.G.; Wikström, H.; Lindberg, P.; Sanchez, D.; Hjorth,
    S.; Carlsson, A.; Paalzow, L.; J. Med. Chem. 1981, 24,
    429.
7. Bolin, J.T.; Filman, D.J.; Matthews, D.A.; Hamlin, R.C.;
    Kraut, J.; J. Biol. Chem. 1982, 257, 13650.
8. Roth, B.; Cheng, C.C.; in "Progress in Medicinal
    Chemistry"; Vol. 19; Ellis, G.P.; West, G.B.; Eds.;
    Elsevier; Amsterdam, 1982; pp. 269-331.
9. Clemens, J.A.; Kornfeld, E.C.; Phebus, L.A.; Shaar, C.J.;
    Smalstig, E.B.; Cassady, J.M.; Nichols, D.E.; Floss, H.G.;
    Kelly, E.; in "Chemical Regulation of Biological
    Mechanisms"; Spec. publ. 42; Creighton, A.M.; Turner, S.;
    Eds.; Royal Society of Chemistry; London, 1982; pp. 167-
    180.
10. Berney, D.; Schuh, K.; Helv. Chim. Acta 1982, 65, 1304.

RECEIVED March 3, 1983

Conformationally Defined Pyrroloisoquinoline Antipsychotics

Implications for the Mode of Interaction of Antipsychotic Drugs with the Dopamine Receptor

G. L. OLSON, H.-C. CHEUNG, E. CHIANG, and L. BERGER

Hoffmann-La Roche Inc., Chemical Research Department, Nutley, NJ 07110

The pyrrolo- and cycloalka [4,5] pyrrolo[2,3-g] isoquinoline ring systems (1) were designed on the basis of a hypothetical model of the interaction of antipsychotic drugs with the dopamine receptor. The prototype (1a)(2,6-dimethyl-3-ethyl-4,4a,5,6,7,8,8a,9-octahydro-4a,8a-trans-1H-pyrrolo [2,3-g] isoquinolin-4-one) (Ro 22-1319) is a potent, selective D-2 dopamine receptor antagonist which exhibits potent antipsychotic-like activity in animal tests and is being evaluated clinically. A series of analogs has been synthesized to probe the effects of substituents and ring size on pharmacological activity and receptor binding. Introducing bulky groups in the 2- and 3-positions, or increasing ring size in the cycloalka analogs diminishes activity and reveals a steric barrier near the 2-position. A wide range of substituents on the basic nitrogen are consistent with pharmacological activity, but only compounds having lipophilic substituents are proportionally potent in ^{3}H-spiroperidol binding. The results suggest that interactions of the nitrogen substituent with the auxiliary binding site identified in the model modulates the activity between D-1 and D-2 dopamine receptors.

Antipsychotic drugs have emerged from a wide variety of structural classes (Figure 1) including tricyclics such as the phenothiazines and thioxanthenes, butyrophenones, diphenylbutylpiperidines, benzamides, indolones and others. Despite this structural diversity, all of these drugs have the common property of acting as postsynaptic dopamine receptor antagonists, although their relative affinity for different subpopulations of receptors may differ.

Structural Requirements

Over the years there has been an evolution in understanding the structural requirements for dopamine antagonist activity and in

0097-6156/83/0224-0251$07.00/0

appreciating the common characteristics of these diverse structures. For example, Janssen (1) developed a linear classification scheme (Figure 2) which indicated the presence of an aromatic moiety separated by a short chain (usually of four atoms) from a basic nitrogen in these compounds. A composite picture (Figure 3) containing features of many known antipsychotics was devised by Gschwend (2), and conveys more structural information.

The advent of dopamine receptor binding assays (3) is a key development that continues to influence the study of structure-activity relationships among antipsychotics. A major insight from binding assays is the recognition that the common property of diverse structural types of both agonist and antagonist drugs is their interaction with closely related macromolecular receptors. Feinberg and Snyder (4) recognized the three-dimensional structural relationship between low energy conformers of dopamine and antagonists of the tricyclic phenothiazene class (e.g. chlorpromazine, Figure 4). Humber (5) built features of the tricyclic antagonists and features of the agonist apomorphine into the relatively rigid and chiral framework of butaclamol (Figure 5). These studies provided the first three-dimensional representations of the pharmacophore. Humber's structure-activity studies (6, 7) on butaclamol and analogs provided a useful abstract representation (Figure 6) of the shape and geometry of a receptor surface.

A Hypothetical Receptor Model

Our approach to understand how the different classes of antipsychotic drugs are related has been to use what is known about the structures and pharmacophoric groups of the drug compounds to infer a hypothetical molecular structure of the receptor binding site. By this approach, we are attempting to treat the binding of drugs to receptors in analogy to small molecule-protein interactions, such as those observed in X-ray crystallographic studies of enzyme-substrate/inhibitor complexes. The dopamine receptor model (8) has been developed by considering plausible intermolecular interactions between functional groups on the drug compounds and complementary amino acid side chain functional groups that might be present on a dopamine receptor protein. From this basis, the three dimensional relationship of these hypothetical receptor groups has been mapped using information from compounds of different structural classes and rigid analogs (8). In this article the basis of the hypothetical model is reviewed briefly and structure-activity relationships from a series of compounds designed and modified on the basis of the model are presented.

In agreement with previous work, the essential pharmacophoric groups are defined by dopamine, dopamine agonists such as (R)-apomorphine, and neuroleptic drugs, which all possess a basic nitrogen atom separated by a 5-7 Å chain or framework from an aromatic ring. It appears reasonable to assume that these two groups are pharmacophores for agonists and antagonists and are bound to complementary functionality on the receptor protein (Figure 7). We imagine the basic amino group of the drug to bind to an acidic amino acid residue (e.g., Asp, Glu; perhaps the

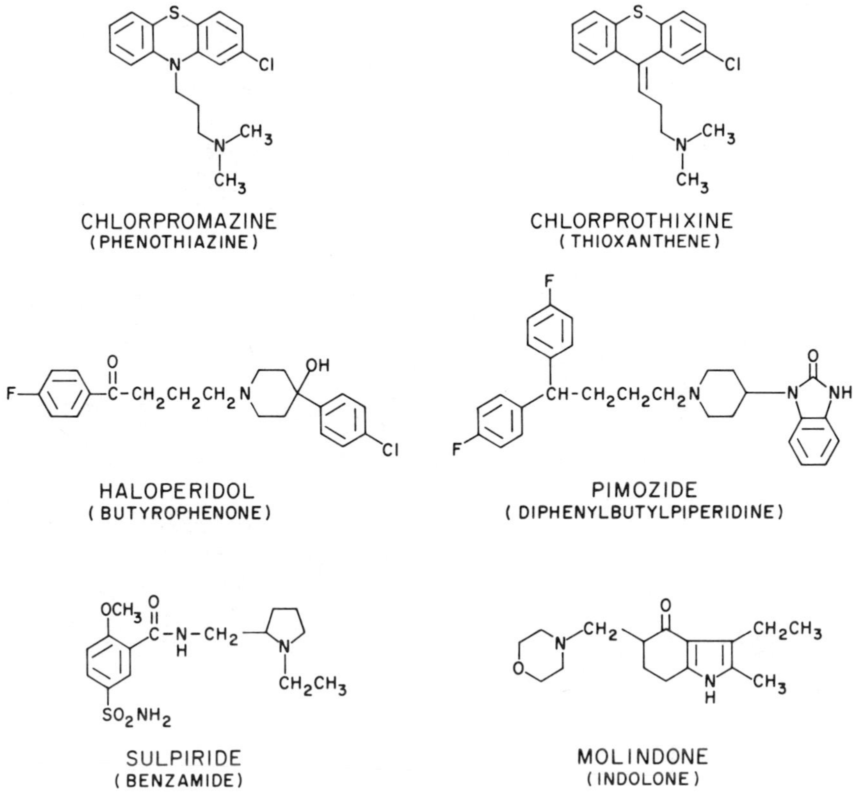

CHLORPROMAZINE
(PHENOTHIAZINE)

CHLORPROTHIXINE
(THIOXANTHENE)

HALOPERIDOL
(BUTYROPHENONE)

PIMOZIDE
(DIPHENYLBUTYLPIPERIDINE)

SULPIRIDE
(BENZAMIDE)

MOLINDONE
(INDOLONE)

Figure 1. Representative antipsychotic drugs.

1. A–C–C–C–C–N–C
2. A–C=C–C–C–N–C
3. A–N–C–C–C–N–C
4. A–O–C–C–C–N–C
5. A–S–C–C–C–N–C
6. A–C–N–C–C–N–C
7. A–C–O–C–C–N–C
8. AA'–C–C–N–C
9. A'–C–C–N–C

Figure 2. Classification scheme for neuroleptics. (Reproduced with permission from Ref. 1. Copyright 1973, Pergamon Press Ltd.)

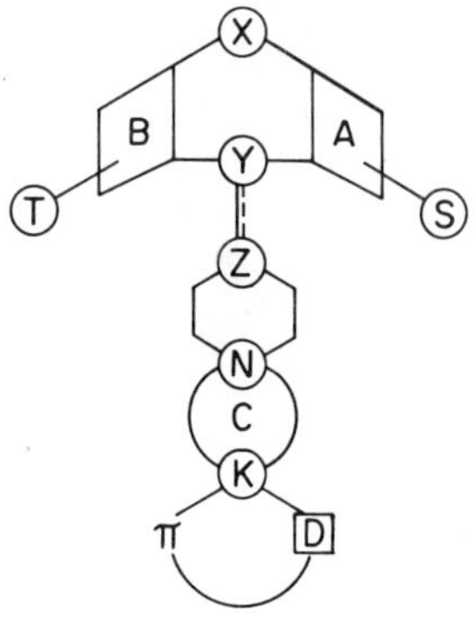

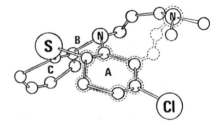

*Figure 3. Composite structure for neu-
roleptics. (Reproduced with permission
from Ref. 2. Copyright 1974, Futura
Publishing Company, Inc.)*

*Figure 4. Relationship between chlor-
promazine and dopamine. (Reproduced
with permission from Ref. 4.)*

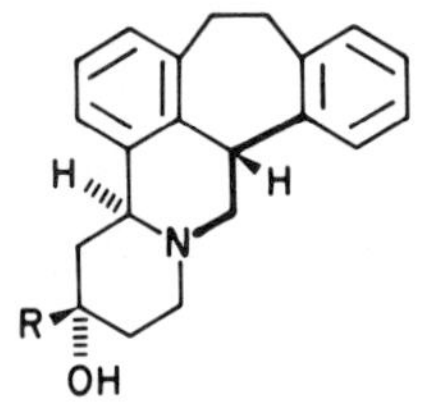

(+) – BUTACLAMOL   (R = C(CH₃)₃)

(+) – DEXCLAMOL   (R = CH(CH₃)₂)

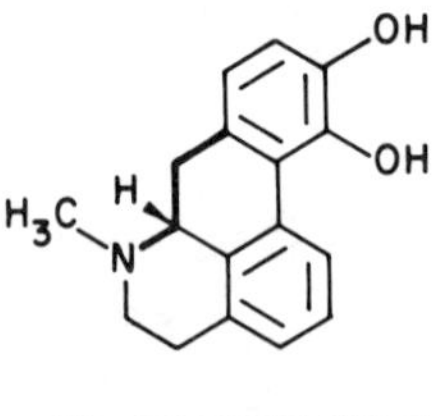

*Figure 5. Relationship between (R)-
apomorphine and (+)-butaclamol/(+)-
dexclamol (5).*

(R̲)-APOMORPHINE

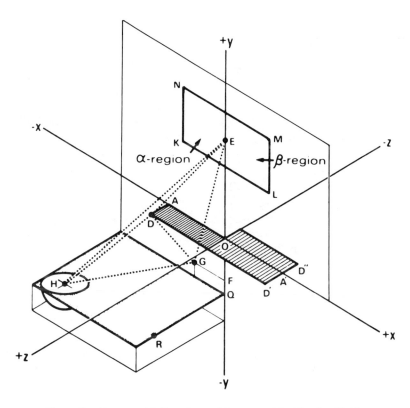

Figure 6. Receptor binding site model proposed by Humber (6, 7).

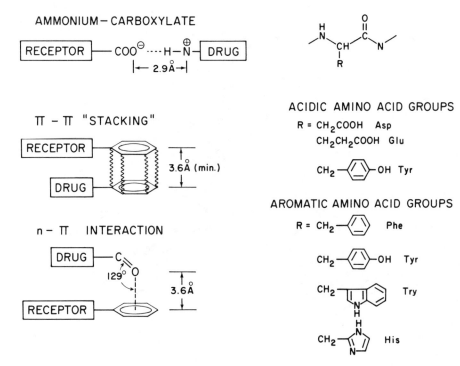

Figure 7. Drug–receptor interactions invoked in hypothetical dopamine receptor model.

phenolic OH of Tyr). We consider the aromatic ring to form a π- π stacking interaction with an aromatic amino acid residue (e.g., Phe, Tyr, Try, His). Also invoked in our model is the interaction of the nonbonded lone pair electrons of a carbonyl oxygen with an aromatic amino acid residue (see below). The plausibility of these interactions has been previously discussed (8). The separation between the receptor features which bind the aromatic ring and the basic nitrogen and the geometrical relationship between them can be defined by considering how representative compounds (especially rigid analogs) could bind. We have used a simple general scheme (Figure 8) to describe the relationship between the aromatic and basic amine pharmacophoric groups independently of the connecting atoms. The data for the torsion angle τ shows the B conformer of (+)-dexclamol to resemble apomorphine more closely than the A conformer found in the crystal of (+)-butaclamol, in agreement with Humber's model (5, 6, 7).

When this model of interaction is applied to other classes of antipsychotic compounds (Figure 9) two other features of the binding site emerge. Thus, in dexclamol, all tricyclic, and diphenylbutylpiperidine compounds, a second aromatic ring appears at a location which could match that occupied by ring C of dexclamol. In butyrophenones, the presence of a ketone carbonyl instead of this second aromatic moiety is apparent. A carbonyl group at this position is also seen in atypical neuroleptics such as molindone and the benzamides. Since the common characteristic of a carbonyl oxygen and an aromatic ring is their localization of electron density, we speculate that a third molecular feature on the receptor may be another aromatic amino acid residue. This residue could bind ring C of dexclamol or the second aromatic ring of the tricyclics or diphenylbutyl-piperidines by a stacking interaction. The carbonyl groups of the other neuroleptics could bind by interaction of the n electrons with the face of the aromatic ring of the receptor. To our knowledge, all dopamine receptor antagonist drugs possess functionality capable of donating electrons in this manner.

This third binding site would not be occupied by functionality on dopamine or rigid congeners such as ADTN, and the second aromatic ring of (R)-apomorphine is directed away from this site. These observations offer a tentative molecular rationalization of the mechanism of antagonism by implying that binding this third site by some electronegative group leads to antagonist activity and is unnecessary for agonist activity.

A fourth binding region of low structural specificity can also be identified (dashed box in Figure 9). It may bind bulky alkyl or aryl groups and a variety of spiropiperidine or benzimidazolone groups seen in the butyrophenone analogs. This site is most probably a lipophilic cavity of large dimension nearest the binding site for the basic nitrogen. Since several pharmacologically active compounds lack lipophilic functionality to occupy this site, it is referred to as an auxiliary binding site. Structure-activity and receptor binding studies (see below) suggest that the auxiliary binding site discriminates D-1 and D-2 dopamine receptor antagonists (9).

Development of a Pyrroloisoquinoline Antipsychotic (Ro 22-1319)

Study of this receptor model led to the design of a series of 4a,8a-trans-pyrrolo [2,3-g] isoquinoline derivatives (e.g., Ro 22-1319, **1a**)

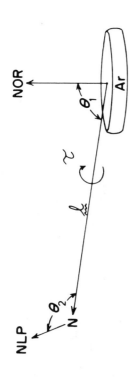

Compound	$|\ell|$, Å	θ_1,deg	θ_2,deg	τ,deg
(R)-apomorphine-1	5.12	99.1	99.4	-9.1
(R)-apomorphine-2	5.09	104.1	101.9	7.7
(+)-dexclamol (conformer B)	4.92	101.3	108.6	45.3
(+)-butaclamol (conformer A)	5.13	84.3	95.5	65.1
pyrroloisoquinoline (-)-1a	5.93	88.6	90.6	-19.1

Figure 8.   Angular relationships of pharmacophores in drugs acting at dopamine receptors (8).

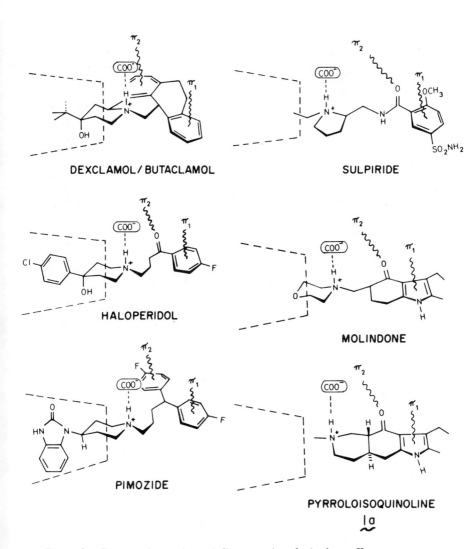

Figure 9.   Receptor interactions of diverse antipsychotic drugs. Key: π_1, π_2, aromatic groups; COO⁻, carboxylate groups; and – – –, auxiliary binding site (8).

incorporating some functionality conceptually derived from molindone and a conformation and configuration built into a rigid molecular framework to conform to the receptor model.

The 4a,8a-trans ring fusion of these compounds assures that the aromatic ring, nitrogen lone pair, and carbonyl group are fixed in an orientation to optimize receptor interaction with the first, second, and third essential binding sites, respectively. The torsion angle τ (Figure 8) between the normal to the aromatic ring and the nitrogen-nitrogen lone-pair bond in these compounds is $-19°$. Thus, in terms of conformation, **1a** resembles apomorphine ($\tau = -9.1°$) more closely than dexclamol (conformer B) ($\tau = 45.3°$). Nevertheless, the extremes in the torsion angle τ ($-19°$ to $+45.3°$) in the antagonist compounds represents only a narrow range ($\pm 32.5°$ centered about $12.5°$) that is easily accommodated by small motions of the receptor (vide infra).

We have modeled the key interactions described with dexclamol (conformer B) and our pyrroloisoquinoline using ORTEP and with space-filling models (Figure 10). The relationship between the pharmacophoric groups in two different structures is usually treated by performing a least-squares superposition of atoms thought to be part of the pharmacophore. As illustrated in Figure 10, this approach has been modified to include the key molecular features of the receptor site in the fitting as they are oriented to bind both dexclamol (conformer B) and **1a**. When the receptor groups are viewed in this superposition in the absence of the drug molecules, the variation in the position of the receptor groups may be considered to represent the extent of motion required by the receptor to bind both drugs optimally. As is evident from the figure, such motions represent only slight changes in the position and orientation of amino acid side chain groups. No gross alteration in the secondary structure of the receptor protein needs to be invoked to understand the adaptation of the receptor to these different drug ligands.

Pharmacology and Biochemistry of Ro 22-1319

The pyrroloisoquinoline derivative Ro 22-1319 (**1a**) has been shown to exhibit potent antipsychotic-like activity in pharmacological tests (10) (Table I). In the avoidance test, Ro 22-1319 is in the potency range of haloperidol and over five times the potency of molindone. The biological activity is highly stereoselective, with virtually all the activity residing in the 4aR,8aR-enantiomer (-)-**1a**. In other aspects of its pharmacology, Ro 22-1319 resembles haloperidol, except that it has relatively weaker cataleptogenic and antistereotypic activity and weaker autonomic effects. It has minimal activity in a rat chronic stereotypy model of receptor supersensitivity, indicating a low potential for tardive dyskinesia. Recent biochemical studies (11) indicate that Ro 22-1319 has sodium-dependent binding properties and has no effect on the dopamine-stimulated adenylate cyclase D-1 receptor which characterize it as a specific D-2 dopamine receptor antagonist like sulpiride, metoclopramide, and molindone (9, 12). Early clinical trials support the conclusions from animal models that Ro 22-1319 is an efficacious antipsychotic agent with minimal extrapyramidal effects.

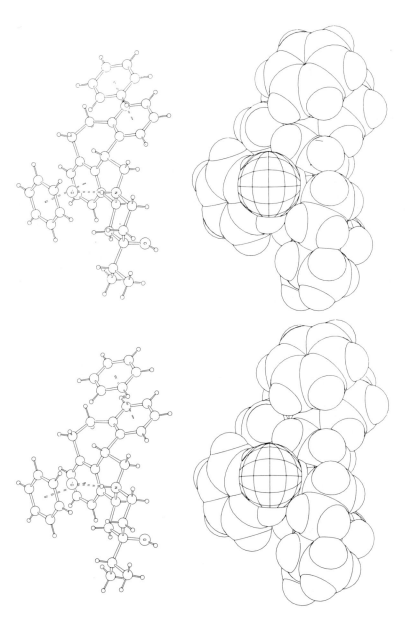

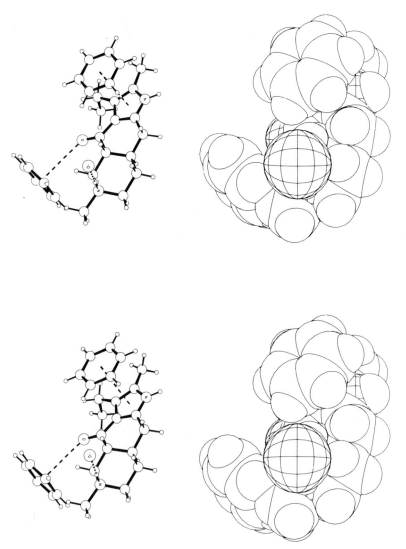

Figure 10b.   Pyrroloisoquinoline (—)-1a and idealized receptor groups.

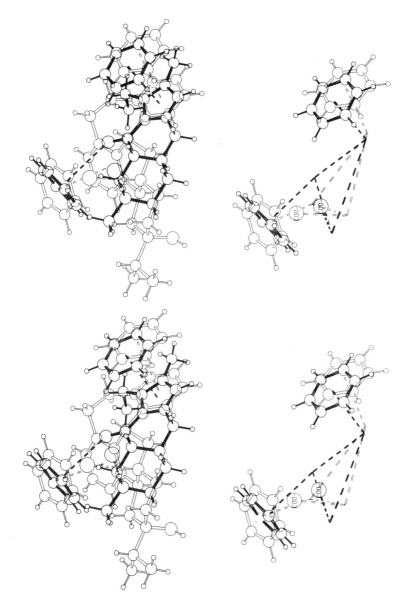

Figure 10c. Least-squares superposition of Figure 10a and Figure 10b which shows the motion of receptor groups required to bind both compounds optimally.

Ia (Ro22-1319)

Table I
Activity of Compounds in the Rat Discrete Avoidance Test

Compound	Avoidance Blockade ABD50 mg/kg p.o.	Escape Blockade EBD20 mg/kg p.o.
(±)-1a·HCl	0.72 (0.43-1.26)[a]	9.8
(-)-1a·HCl	0.49 (0.18-1.22)	3.6
(+)-1a·HCl	6.0	b
haloperidol	0.35 (0.17-0.66)	2.7
molindone	3.6 (0.5-10.5)	17.1

[a] Figures in parentheses are 95% confidence limits.
[b] Not applicable because of lack of avoidance blockade.

Synthesis and Structure-Activity Relationships

The hypothetical model used in the design of Ro 22-1319 has proved to be of considerable value in designing analogs and in interpreting structure-activity relationships. We have been particularly interested in preparing specific compounds to favor or disfavor certain interactions with the essential binding sites and to explore the characteristics of the auxiliary binding site.

The compounds were synthesized (Scheme I) via a Knorr condensation of a 2-isonitrosoketone (ii) and Zn or an aminoketone (iii) with the N-methyloctahydroisoquinolinedione (i) . A new, improved route to the diketone (i) developed by Drs. D. Coffen and U. Hengartner has contributed significantly to the project. Demethylation of the initial product (iv) using ethyl chloroformate followed by alkaline hydrolysis gave the secondary amine (v) which was alkylated with a variety of alkylating agents to give a range of analogs (1). The major products from both the Knorr condensation and the alkylation steps were 4a,8a-trans isomers; these were easily separated from the minor 4a,8a-cis compounds by chromatography or crystallization.

Pyrroloisoquinoline analogs having substituents in the 2- and 3-positions (Table II) ranging in size from H to isopropyl and phenyl were prepared. The unsubstituted compound (2) was relatively weak with dialkyl substitution by sterically non-demanding groups (4, 1a, 5) giving the most potency. Isopropyl substitution in the 3-position (6) decreases activity somewhat relative to the n-propyl group (5) , while isopropyl in the 2-position (7) decreases activity more, as does a phenyl group in the 3-position (8). The phenyl group certainly would interfere with a stacking interaction. The greater sensitivity of the 2-position to bulky groups may reflect both an effect on stacking and the possible presence of a steric contact or barrier near that group. The importance of this barrier can also be seen in the cyclic analogs.

As ring size was increased in the cycloalka[4,5] pyrroloisoquinolines (Table III), there was a steady decrease in potency from the 5- to 6- to 7-membered analogs (9, 10, 11). The substituent methylenes adjacent to the pyrrole ring would be expected to lie roughly in the plane of the pyrrole ring in these analogs, so the decreased potency seems less likely to be due to interference with stacking than to steric interference of the methylenes further from the pyrrole with the barrier mentioned previously near the 2-position. The return of activity in the 8-membered analog (12) can be attributed to puckering of the ring to avoid the interaction with the barrier.

The Auxiliary Binding Site and Selectivity for D-2 and D-1 Dopamine Receptors

A large number of analogs bearing lipophilic substituents on the basic nitrogen have been prepared (Table IV) to probe the characteristics of the auxiliary binding site. As is evident from the table, all of the analogs

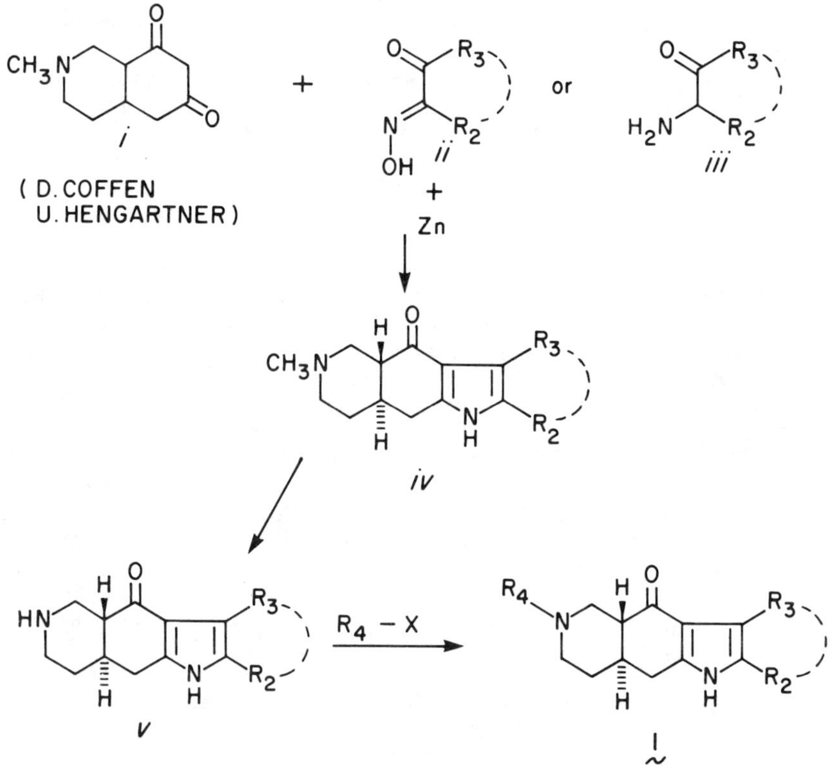

Scheme I. Synthesis of pyrroloisoquinoline and cycloalkapyrroloisoquinoline analogs.

Table II. Effect of substituent bulk at 2- and 3-positions on activity.

COMPOUND	R_2	R_3	AVOIDANCE ABD50 mg/kg po
2	H	H	4.12
3	CH_3	H	1.98
4	CH_3	CH_3	0.47
1a	CH_3	CH_2CH_3	0.7
5	CH_3	$CH_2CH_2CH_3$	0.46
6	CH_3	$CH(CH_3)_2$	1.22
7	$CH(CH_3)_2$	CH_3	35
8	CH_3	(phenyl)	8 po = 10%

Table III. Effect of ring size in cycloalka(4,5)pyrrolo(2,3-g)-
isoquinolines on activity.

COMPOUND	RING	n	AVOIDANCE ABD50 mg/kg po
9		3	0.98
10		4	5.50
11		5	50% AB at 16 po
12		6	7.3

Table IV. Effect of lipophilic groups at basic nitrogen on activity.

COMPOUND	R_4	AVOIDANCE ABD50 mg/kg po	^{3}H-SPIROPERIDOL BINDING IC50 nM
13	H	19.5	195
1a	CH_3	0.7	45.5
14	CH_2CH_3	—	33.5
15	$CH_2CH=CH_2$	2.3	23
16	$CH_2CH_2CH_3$	—	5.6
17	CH_2-◁	1.64	6.9
18	$CH_2CH(CH_3)_2$	—	4.7
19	$CH_2CH_2OCH_2CH_3$	0.8	6.2
20	CH_2-⬡	1.64	0.83
21	CH_2CH_2-⬡	0.37	1.80
22	CH_2CH_2-furyl(O)	1.27	4.7
23	CH_2CH_2-thienyl(S)	0.90	2.4
24	$CH_2CH_2OCH_2$-⬡	1.65	1.25
25	$CH_2CH_2CH_2O$-⬡	0.48	0.58
26	$CH_2CH_2CH(-⬡)_2$	40% at 8 po	6.2
27	$CH_2CH_2CH_2CH(-⬡)_2$	26% at 8 po	3.1

which bear a lipophilic group that could extend into the auxiliary binding site (16–27) are potent in displacing ^{3}H-spiroperidol in vitro, and except for the very insoluble diphenylpropyl (26) and diphenylbutyl (27) compounds, are all potent in avoidance. As shown, the N-methyl compound 1a (Ro 22-1319) does not have a group that could occupy the auxiliary binding site in our model and is relatively weak in the ^{3}H-spiroperidol binding assay, despite having potent in vivo activity. Ro 22-1319 also does not bind in the dopamine-sensitive adenylate cyclase D-1 receptor assay ($IC_{50} > 1000 \mu M$), whereas the N-benzyl analog 17 is a potent D-1 antagonist ($IC_{50} = 0.43 \mu M$).

Recently, Nakamura and Kuruma (11) have demonstrated that ^{3}H-Ro 22-1319 binding is critically dependent on the presence of sodium ion; in this respect resembling ^{3}H-sulpiride binding, and contrasting with ^{3}H-spiroperidol binding, which is sodium independent. In addition, like D-2 antagonistic antipsychotics, Ro 22-1319 interacts with striatal ^{3}H-spiroperidol binding sites in a sodium-dependent manner (11). On this basis, Ro 22-1319 is characterized as a specific D-2 dopamine receptor antagonist.

In our model, we have indicated that atypical antipsychotics (12) (sulpiride, metoclopramide, molindone, and Ro 22-1319) differ from classical neuroleptics (tricyclics, butyrophenones, butaclamol, diphenylpiperidines) by lacking a lipophilic functional group on the basic nitrogen that could extend into the auxiliary binding site identified in our model. The absence of this lipophilic functionality may now be stated to be the characteristic which distinguishes selective D-2 dopamine receptor antagonists from non-selective antagonists.

A comparison of in vivo activity and ^{3}H-spiroperidol binding in the pyrroloisoquinoline series provides an indication of the lipophilicity required on the nitrogen substituent to retain D-2 receptor selectivity. Among the analogs listed in Table IV, there appears to be a break between compounds having alkyl substituents smaller than propyl in the ^{3}H-spiroperidol binding assay ($IC_{50} > 20$ nM) and those with more lipophilic groups (propyl or longer alkyl, various aralkyl) with $IC_{50} < 10$ nM. Further studies are in progress to evaluate the hypothesis that this break corresponds to the transition between selective D-2 antagonism and non-specific (D-1 + D-2) antagonism.

The transition between D-2 and D-1 dopamine receptor binding associated with changes in lipophilicity of a basic nitrogen substituent has also been observed for the metoclopromide analogs YM-09151-2 and YM-08050 (13, 14). Similarly, in the tetrahydroindolone series, replacing the morpholine ring of molindone with a 4-hydroxy-4-phenylpiperidine (15) leads to a dramatic enhancement of ^{3}H-spiroperidol binding potency ($IC_{50} = 0.54$ nM vs $IC_{50} = 74$ nM for molindone) (16).

The character of the auxiliary binding site has been explored further by preparing a series of analogs bearing hydrophilic OH groups in the side chain (Table V). These groups would be expected to interfere with hydrophobic interactions. The trends are clear in that these hydroxyalkyl analogs are weak in the spiroperidol binding assay and yet exhibit potent in vivo activity in avoidance. Our prediction is that these compounds are also specific D-2 receptor antagonists.

Table V.   Effect of hydrophilic groups in side chain at basic nitrogen on activity.

COMPOUND	R_4	AVOIDANCE ABD50 mg/kg po	^{3}H- SPIROPERIDOL BINDING IC50 nM
I a	CH_3	0.7	45.5
28	CH_2CH_2OH	2.5	90
29	$CH_2\overset{OH}{\underset{\mid}{CH}}$—⟨◯⟩	1.7	96
30	$CH_2\overset{OH}{\underset{\mid}{CH}}$-$C(CH_3)_3$	0.62	41

Some of the pharmacologically most potent derivatives in our series of pyrroloisoquinoline and cycloalkapyrroloisoquinolines are those substituted with fluorophenylacyl groups (Table VI). To show that this potency is not the result of making a haloperidol-type butyrophenone out of our pyrroloisoquinoline, the carbonyl group in **31** was reduced to the alcohol **(32)** and the methylene chain shortened (compare **33 - 35**). In contrast to the haloperidol-type compounds, in all cases the alcohols are in the range of potency of the ketones, and shortening the chain by one carbon has only a moderate effect on potency. The binding characteristics of compounds in this series and their activity in tests indicative of side effects are being studied further.

Table VI.   Fluorophenylacyl and fluorophenylhydroxyalkyl analogs.

COMPOUND	X	n	R_2	R_3	AVOIDANCE ABD50 mg/kg po
31	O	3	CH_3	CH_3	0.08
32	H,OH	3	CH_3	CH_3	0.15
33	O	3	CH_3	CH_2CH_3	0.20
34	O	2	CH_3	CH_2CH_3	0.85
35	O	1	CH_3	CH_2CH_3	8 po = 13%
36	O	3	—$(CH_2)_3$—		0.15
37	H,OH	3	—$(CH_2)_3$—		0.31
38	O	3	—$(CH_2)_4$—		0.73
39	H,OH	3	—$(CH_2)_4$—		1.32
40	O	3	—$(CH_2)_6$—		ca. 11

Summary

The hypothetical model we have described for the interaction of antipsychotic drugs with the dopamine receptor has been of significant value in the design and rational modification of a series of novel, highly active pyrrolo- and cycloalkapyrroloisoquinoline antipsychotics. One member of the series, Ro 22-1319, is currently in clinical trials as a therapeutic agent for the treatment of psychosis. A series of analogs has been synthesized to probe features of the dopamine receptor suggested by the model. The structure–activity relationships have illuminated molecular characteristics which differentiate selective D-2 dopamine receptor antagonists from non-selective antagonists.

Acknowledgments

We gratefully acknowledge the contributions to this work of Dr. John F. Blount and Mr. Louis Todaro (X-ray results and ORTEP figures), Dr. Arnold B. Davidson and Mr. Edward Boff (pharmacology and clinical results), and Dr. Robert O'Brien and Mr. Gordon Bautz (^{3}H-spiroperidol binding results), and Drs. David Coffen and Urs Hengartner (process research); all of Hoffmann-La Roche, Nutley. We especially thank Drs. K. Nakamura and I. Kuruma (^{3}H-Ro 22-1319 binding results) of the Nippon Roche Research Center, Kamakura.

Literature Cited

1. Janssen, P.A.J. in "International Encyclopedia of Pharmacology and Therapeutics, Section 5 – Structure–Activity Relationships;" Pergamon Press: New York, 1973; p. 37.
2. Gschwend, H. W. in "Neuroleptics"; Fielding, S.; Lal, H., Eds. Futura: Mt. Kisco, 1974; p. 1.
3. Creese, I.; Stewart, K.; Snyder, S. H. Science 1978, 192, 481.
4. Feinberg, A. P.; Snyder, S. H. Proc. Natl. Acad. Sci. USA 1975, 72, 1899.
5. Humber, L. G.; Bruderlein, F. T.; Voith, K. Mol. Pharmacol. 1975, 11, 833.
6. Humber, L. G.; Bruderlein, F. T.; Philipp, A. H.; Gotz, M.; Voith, K. J. Med. Chem. 1979, 22, 761.
7. Philipp, A. H.; Humber, L. G.; Voith, K. J. Med. Chem. 1979, 22, 768.
8. Olson, G. L.; Cheung, H.-C.; Morgan, K. D.; Blount, J. F.; Todaro, L.; Berger, L.; Davidson, A. B.; Boff, E. J. Med. Chem. 1981, 24 1026.
9. Kebabian, J. W.; Calne, D. B. Nature 1979, 277, 93.
10. Davidson, A. B.; Boff, E.; MacNeil, D.; Wenger, J.; Cook, L. Psychopharmacology, in press.
11. Nakamura, K.; Kuruma, I. personal communication.
12. Rosenfeld, M. R.; Dvorkin, B.; Klein, P. N.; Makman, M. H. Brain Research 1982, 205.

13.   Iwanami, S.; Takashima, M.; Hirata, Y.; Hasegawa, O.; Usuda, S. J.
      Med. Chem. 1981, 24, 1224.
14.   Usuda, S.; Nishikori, K.; Noshiro, O.; Maeno, H. Psychopharmacology
      1981, 73, 103.
15.   Schoen, K.; Pachter, I.; S. Africa, 67 04, 863, Feb. 14, 1968 Chem.
      Abstr. 1969, 71, 30356d.
16.   O'Brien, R. A.; Bautz, G. personal communication.

RECEIVED February 16, 1983

Renal Vascular Dopamine Receptor Topography

Structure–Activity Relationships That Suggest the Presence of a Ceiling

PAUL W. ERHARDT

American Critical Care, McGaw Park, IL 60085

A topographical model has been proposed to explain why (E)-2-(3,4-dihydroxyphenyl)cyclopropylamine, 1, and alpha-methyldopamine (AMDA) are inactive in the renal vascular dopamine (DA) receptor system. In this model a steric protrusion (S2) resides approximately 2Å above the generalized plane of the receptor and acts to impede interaction with molecules such as 1 and AMDA which possess additional bulk in this region. Recent developments in DA structure-activity relationships offer further support for the existence of the S2 site.

The close structural similarity of the semirigid conformational analogue 1 to the trans-rotamer of dopamine (DA) was noted almost ten years ago (1,2). The synthesis of 1 has been described more recently and its pharmacological profile determined (3-6) to be similar to alpha-methyldopamine (AMDA) which is inactive as a dopaminergic agent.

0097–6156/83/0224–0275$06.00/0

It has been suggested (7) that AMDA lacks dopaminergic activity because its aryl-ring cannot achieve approximate coplanarity with its side chain as required (8,9,10) for optimal interaction with DA receptors. However, in AMDA the side chain carbon bonds, and in 1 the bond between the beta-carbon and catechol, are all free to rotate with only minimal thermodynamic constraint (11). Therefore, conformational arguments seem an unlikely explanation for the inactivity observed for these compounds. Alternatively, it has been suggested that the molecular bulk added by the methylene in 1 and the methyl in AMDA interferes with receptor interaction because a steric protrusion resides above the general plane of the receptor in a proposed (12) topographical model.

Examination of considerable DA agonist structure-activity relationships (SAR), with specific consideration of the central chain region, revealed that while aryl-amine near coplanarity is a feature common to all of these agents, the central chain varies considerably from the common plane relationship. This became an important distinction from the prior art since in the receptor model there is then no requirement for agonists to approach an antiperiplanar conformational arrangement or to achieve an approximate 180° dihedral angle. For example, certain indanamine derivatives (1,8) possess aryl-amine coplanarity when assuming an envelope conformation (13), and certain benzazepine derivatives (14) can assume a favored (15) partially eclipsed conformation in which the amine function lies within the plane established by the catechol ring.

Final topographical features for this receptor model were provided by aporphine SAR. It has been suggested (16) that the A-ring of apomorphine (APO) associates with an auxiliary hydrophobic binding site. Alternatively, to explain the inactivity of isoapomorphine (ISO) it has been suggested (16,17, 18) that this same ring resides in a receptor region limited by a steric boundary. The difference between these compounds is that in APO the A-ring is on the same side as the analogous DA meta-hydroxy group while in ISO this ring is opposite to the analogous meta-hydroxy group. This difference becomes apparent when an attempt is made to place APO and ISO in the receptor model as depicted in Figure 1.

Examination of recent DA SAR further supports the existence of an S2 site. Several new compounds, 2-7, contain steric bulk in regions nearly superimposable with that in 1 and AMDA, as illustrated by the arrow next to their Newman projections. Although many of these structures are held in putative ideal conformations, their indicated bulk would be expected to collide with site S2 in Figure 1. All of these structures have been determined (19-24) to be essentially inactive in dopaminergic test systems. To date, this author has not found a potent peripheral or central DA agonist structure which manifests significant bulk in the receptor region designated as S2.

2(19)

3(20)

4(21)

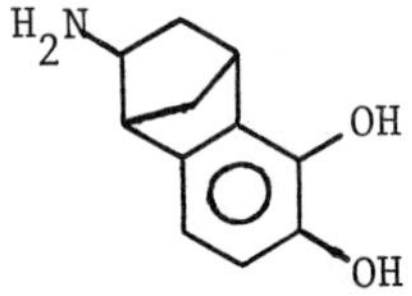

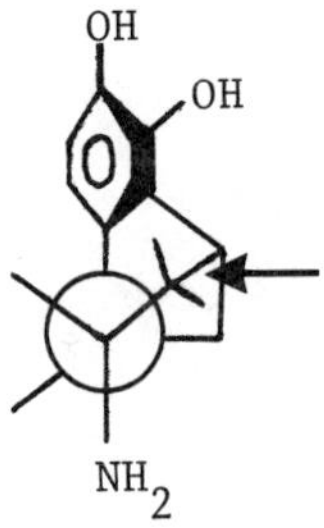

5 (<u>22</u>)

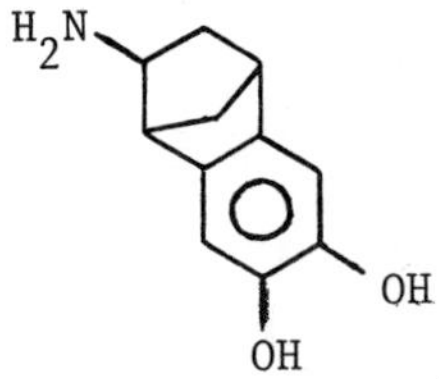

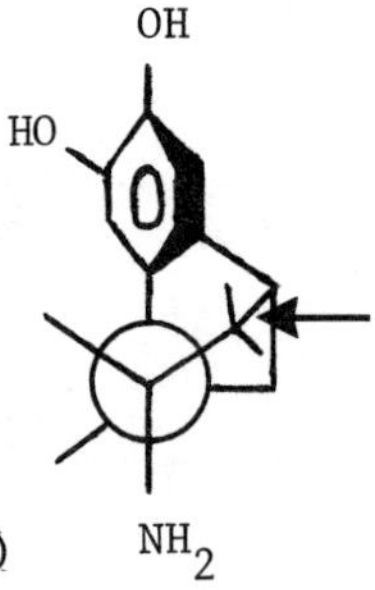

6 (<u>22</u>, <u>23</u>)

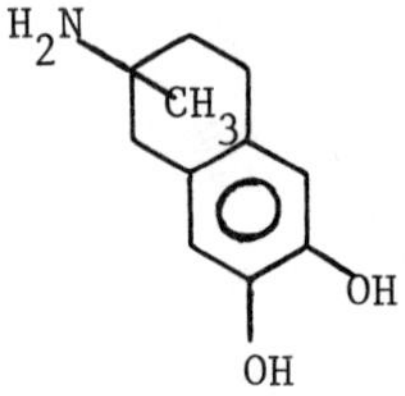

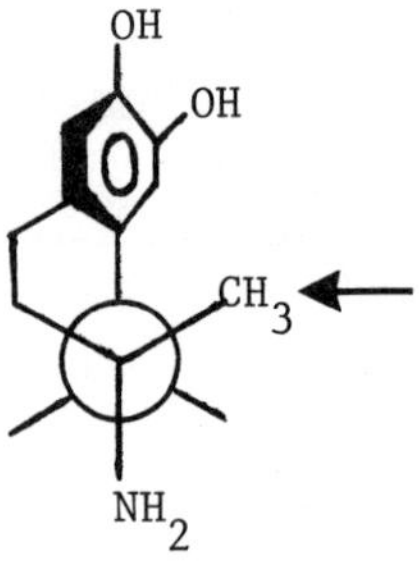

7 (<u>24</u>)

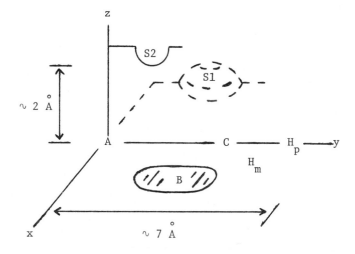

Figure 1. Topographical model of the renal vascular dopamine receptor (12).
Features include: 1, a single plane containing amine (A) and catechol (C, H_m, H_p) recognition sites (the distances between A and H_m and A and H_p approximate 7 Å; the plane extends across the analogous central chain region and can be thought of as the "floor" of the receptor); 2, a steric parameter, S1 ("rear wall"), located in the −x, +y quadrant and a steric parameter, S2 ("ceiling"), centered specifically near x, 0.0; y, 1.5; z, 2.5; and 3, an auxiliary hydrophobic or lipophilic binding site (B) located in the +x, +y quadrant. With refinement, Site A can be moved to a more specific location approximately 0.2 Å above the x–y plane and the nature of Site C need not be specifically defined as a recognition or binding site. The rigid representation in this model is not meant to imply that a more dynamic relationship, such as mutual molding, is not operative during drug–receptor interaction. (Reproduced with permission from Ref. 12. Copyright 1980, American Pharmaceutical Association.)

Literature Cited

1. Pinder, R.M. Adv. Neurol. 1973, 3, 295.
2. Costall, B.; Naylor, R.J.; Pinder, R.M. J. Pharm. Pharmacol. 1974, 26, 753.
3. Erhardt, P.W. J. Org. Chem. 1979, 44, 883.
4. Erhardt, P.W.; Gorczynski, R.J.; Anderson, W.G. J. Med. Chem. 1979, 22, 907.
5. Gorczynski, R.J.; Anderson, W.G.; Erhardt, P.W.; Stout, D.M. J. Pharmacol. Exp. Ther. 1979, 210, 252.
6. Borgman, R.J.; Erhardt, P.W.; Gorczynski, R.J.; Anderson, W. G. J. Pharm. Pharmacol. 1978, 30, 193.
7. Cannon, J.G.; Perey, Z.; Long, J.P.; Rusterholz, D.B.; Flynn, J.R.; Costall, B.; Fortune, D.H.; Naylor, R.J. J. Med. Chem. 1979, 22, 901.
8. Cannon, J.G. Adv. Neurol. 1975, 9, 177.
9. Cannon, J.G.; Smith, R.V.; Aleem, M.A.; Long, J.P. J. Med. Chem. 1975, 18, 108.
10. Cannon, J.G.; Hatheway, G.J.; Long, J.P.; Sharabi, F.M. J. Med. Chem. 1976, 19, 987.
11. Pullman, B.; Coubeils, J.L.; Courriere, P.H.; Gervois, J.P. J. Med. Chem. 1972, 15, 17.
12. Erhardt, P.W. J. Pharm. Sci. 1980, 69, 1059.
13. Fuchs, B. "Topics in Stereochemistry" (ed. Eliel, D.; Allinger, N.), Vol. 10, Wiley, New York, 1978.
14. Pendleton, R.G.; Samler, L.; Kaiser, C.; Ridley, P.T. Eur. J. Pharmacol. 1978, 51, 19.
15. Sauriol-Lord, F.; Grindley, T.B. J. Am. Chem. Soc. 1981, 103, 936.
16. Goldberg, L.I.; Kohli, J.D.; Kotake, A.N.; Volkman, P.H. Fed. Proc. Fed. Am. Soc. Exp. Biol. 1978, 37, 2396.
17. Grol, C.J.; Rollema, H. J. Pharm. Pharmacol. 1977, 29, 153.
18. McDermed, J.D.; Freeman, H.S.; Ferris, R.M. "Catecholamines: Basic and Clinical Frontiers" (ed. Usdin, E.; Kopin, I.J.; Barchas, J.), Vol. 1, Pergamon, New York 1978, p. 568.
19. Cannon, J.G.; Hatheway, G.J.; Long, J.P.; Sharabi, F.M. J. Med. Chem. 1976, 19, 987.
20. Cannon, J.G.; Lee, T.; Hsu, F.-L.; Long, J.P.; Flynn, J.R. J. Med. Chem. 1980, 23, 502.
21. Law, S.-J.; Morgan, J.M.; Masten, L.W.; Borne, R.F.; Arana, G.W.; Kula, N.S.; Baldessarini, R.J. J. Med. Chem. 1982, 25, 213.
22. Burn, P.; Crooks, P.A.; Heatley, F.; Costall, B.; Naylor, R.; Nohria, V. J. Med. Chem. 1982, 25, 363.
23. Schuster, D.I.; Katerinopoulos, H.E.; Holden, W.L.; Narula, A.P.S.; Libes, R.B. J. Med. Chem. 1982, 25, 850.
24. Nichols, D.E.; personal communication.

RECEIVED February 16, 1983

INDEX

Jacket design by Kathleen Schaner
Indexing and production by Deborah Corson and Frances Reed

Elements typeset by Service Composition Co., Baltimore, MD
Printed and bound by Maple Press Co., York, PA